NGOs as Advocates for Development in a Globalising World

To maximise their impact, NGOs have globalised their operations and formed new strategic alliances to increase the effectiveness of their advocacy. Historically, as a means of improving the life of people in disadvantaged communities, Northern development NGOs primarily acted at the local scale in developing nations. In the last decade, these NGOs have increasingly given resources to advocacy campaigns directed at global and regional actors, including multilateral banks, government and corporations. To date, there has been little basis for gauging the extent to which NGOs' advocacy work has contributed to poverty alleviation.

This book traces this recent growth in NGO advocacy. Rugendyke presents empirical findings about the impacts of NGO advocacy activity on the policies and practices of global and regional institutions. Case studies illustrate the advocacy work of Australian NGOs, of British NGOs' policies about engaging with multinationals, of Oxfam's advocacy directed at World Bank policies and NGO advocacy in the Mekong Region.

Adopting an interdisciplinary approach, this book examines the mixed successes of advocacy as a strategy used by NGOs in attempting to address the ongoing causes of poverty in developing nations. It will be a useful aid to researchers, students and lecturers and to development practitioners interested in advocacy as a development strategy.

Barbara Rugendyke is a Senior Lecturer in Geography at the University of New England, Armidale, Australia.

NGOs as Advocates for Development in a Globalising World

Edited by Barbara Rugendyke

Routledge
Taylor & Francis Group

LONDON AND NEW YORK

First published 2007
by Routledge
2 Park Square, Milton Park, Abingdon, Oxon OX14 4RN

Simultaneously published in the USA and Canada
by Routledge
270 Madison Ave, New York, NY 10016

Routledge is an imprint of the Taylor & Francis Group, an informa business

© 2007 Barbara Rugendyke

Typeset in Times New Roman by
Keystroke, 28 High Street, Tettenhall, Wolverhampton
Printed and bound by
Antony Rowe Ltd, Chippenham, Wiltshire

British Library Cataloguing in Publication Data
A catalogue record for this book is available from the British Library

Library of Congress Cataloging in Publication Data
A catalog record for this book has been requested

ISBN 10: 0–415–39530–5 (hbk)
ISBN 10: 0–415–39531–3 (pbk)
ISBN 10: 0–203–93921–2 (ebk)

ISBN 13: 978–0-415–39530–4 (hbk)
ISBN 13: 978–0-415–39531–1 (pbk)
ISBN 13: 978–0-203–93921–5 (ebk)

To Anya and David

Overcoming poverty is not a gesture of charity. It is an act of justice. . . .

Recognise that the world is hungry for action, not words. Act with courage and vision.

. . . Sometimes it falls on a generation to be great. You can be that great generation.

(Nelson Mandela, Trafalgar Square, 3 February 2005)

Contents

Figures

Tables

Contributors

Ian Anderson, a professional accountant, had a successful career as a specialist tax partner in a major accounting firm and later as a senior banker working in structured finance in Australia and Hong Kong. During 20 years in Hong Kong, Ian continued his engagement in development, humanitarian relief and global equity issues, which began with Community Aid Abroad in his native Melbourne. Ian joined the Standing Committee of the Hong Kong Oxfam Group in 1981, and was founding Chair of Oxfam Hong Kong, serving from 1988 until 1997, and thereafter Vice-Chair and later a Council member until 2003. Ian was a founder board member of Oxfam International and elected Chair of the Board of Oxfam International from 1999 until 2003. Returning to Australia, Ian resumed his engagement with Oxfam Community Aid Abroad, now Oxfam Australia, and was appointed to its board in 2003. Ian also served as Chair of Australians for Just Refugee Programs Inc. and its 'A Just Australia' national campaign related to asylum seekers and refugee policy. Ian's doctoral thesis, completed through the University of New England in Australia, examined international NGO advocacy through Oxfam International's influence on the World Bank's poverty reduction policy.

Philip Hirsch is Associate Professor of Geography at the University of Sydney and has published extensively on rural development, natural resource management and the politics of the environment in South-East Asia, most notably on Thailand. His publications include work on livelihood-oriented NGOs and on advocacy that employs environment as a legitimating discourse for wider livelihood struggles. He is the Director of the Australian Mekong Resource Centre, which works closely with both academic institutions and advocacy networks in Australia and countries of the Mekong Region. His academic training in both geography and social anthropology gives him a combination of village based and global perspectives on development. His collaborative research work with academic, NGO, local government and international institutional partners in the Mekong Region over many years gives him an inside understanding of NGO–government dynamics in several Mekong countries.

Cathryn Ollif travelled the 'hippie trail' from London to Kathmandu as a teenager, including Iran, Iraq and Afghanistan, and there her interest in development issues emerged. This interest led her to New York where, at 22, she began to work with the United Nations Organization. Ten years of employment with the UN included six years in the Secretariat in New York and four years in peacekeeping, based in Cyprus and visiting missions in Syria, Lebanon, Israel and Egypt. Returning to Australia after 16 years, Cathryn was led by her interest in global processes to engage in tertiary study which focused on development education and included field work in India with Oxfam Community Aid Abroad. This was followed by successful completion of a doctoral thesis, through the University of New England in Australia, which examined the advocacy work of Australian development NGOs. Cathryn has recently become involved in human rights activism related to asylum seekers in Australia, which has contributed to her understanding of the broader issues surrounding NGO advocacy work. Cathryn is currently employed as a Parliamentary Secretary in Canberra, Australia.

Barbara Rugendyke's doctoral research focused on the policies and practices of Australian non-government development assistance agencies. This interest led to involvement as a board member of Australian development NGOs and related research and publication. Now a Senior Lecturer in Geography, Barbara has been teaching about development issues at the University of New England in Australia, for over a decade. Barbara has supervised research students studying a wide range of development-related issues in many countries. Her own research has focused on community development planning in remote indigenous communities, the social and environmental impacts of tourism, particularly in Vietnam, and the activities of development NGOs.

John Sayer was a farmer in Wales before moving to Asia in 1976. He worked on rural development projects in India before settling in Hong Kong, where he lived for 25 years. He joined the Asia Monitor Resource Centre (AMRC) to research the impact of foreign economic involvement on Asian economies and workers. As Executive Director of AMRC he worked with trade unions and NGOs in Asia on research and information systems, computer communications and occupational health and safety. In 1991 he joined Oxfam Hong Kong as Program Director and later became Executive Director, working on relief and development in many of the poorest nations of Asia and Africa. In 2001 he moved to Oxford to take up the post of interim Executive Director of Oxfam International. While working as a consultant to Oxfam on relations between NGOs and the private sector, he completed postgraduate research through the University of New England in Australia, writing about NGO engagement with the corporate sector. He then served as Director of Africa Now, a development agency assisting poor producers in Africa through small enterprise development, equitable market access and the promotion of ethical trade. In 2006, he again moved to Hong Kong, where he now serves as Executive Director of Oxfam Hong Kong.

Lindsay Soutar spent the last five years travelling between Australia and the countries of the Mekong Region. In her work with the Australian Mekong Research Centre she has carried out research on topics including Australian aid policy, regional development policy, and teaching and learning methods for environmental education. She recently spent a year in Thailand working with the Mekong Learning Initiative, a collaborative network promoting innovative approaches to environmental education in the Mekong Region. Lindsay completed a Bachelor of Economics in Political Science and Geography at the University of Sydney in Australia. Her First Class Honours research in geography drew on her undergraduate studies in both geography and political science in addressing the relationship between the Asian Development Bank and NGOs in the Mekong Region.

Acknowledgements

A belief that a more just world is worth striving for drives the authors who have contributed to this volume. They share a belief that by working alone, but perhaps better through networks, associations and organisations, individuals can contribute to positive change. However, none accept uncritically the work currently undertaken by non-government development organisations (NGOs) in seeking to eradicate poverty and disadvantage, the daily reality faced by far too many in our world. To be effective catalysts of change, these organisations must regularly reassess their reason for being and their ability to fulfil their mandate. This collection presents substantial empirical research about the extent and impacts of, and challenges facing, development NGOs' attempts to tackle global poverty through increased advocacy. If, in some small way, it enhances understanding of that increasingly important global phenomenon, then it has achieved its purpose.

Working with enthusiastic researchers with shared interests, who are committed to further understanding the effectiveness of NGOs in improving the lives of those on whose behalf they advocate, has been a great privilege. The willingness of Cathryn Ollif, Ian Anderson and John Sayer to contribute some of their detailed research about the advocacy of development NGOs to this volume made completion of the project a possibility. While their research provides the substance of much of this book, contributions by Lindsay Soutar and Philip Hirsch broaden its focus, enabling some understanding of the intricacies of NGO advocacy in an important and complex part of the world: the Mekong Region, where political space has usually been severely circumscribed. I am grateful to them all for supporting this project, which aims to bring together recent empirical research about the advocacy activities of NGOs in pursuit of their mission to reduce global poverty.

It is impossible to thank here the countless numbers of people who have informed the research presented by the authors in this volume. We have all learnt from the NGO staff, staff of bilateral and multilateral agencies, politicians and media workers who have generously shared of their time, resources and ideas. Without the assistance and shared wisdom of those many willing participants in the process, the research reported in these pages could not have materialised. Although most have elected to remain anonymous and cannot be named here, to them we are grateful.

Many individuals have kindly provided permission to use material published elsewhere and are gratefully acknowledged here. Oxfam United Kingdom helpfully sourced the cover image, photographed by Toby Adamson, and provided permission for its use. Allen & Unwin granted permission for reuse of some material in Chapter 1, previously published as B. Rugendyke (1991) 'Unity in diversity: the changing face of the Australian NGO community', in L. Zivetz *et al.*, *Doing Good: The Australian NGO Community*. Sheila Smith of UNA Exchange kindly allowed the use of Figure 1.1, and Lyndsey Maiden of Maiden Photography, Wales, the use of Figure 1.2. Luke Fletcher, National Coordinator of Jubilee Australia, granted permission for the use of Figure 3.1; Oxfam International for Figure 4.1; Philip Hirsch for the photographs in Chapter 8; Towards Ecological Recovery and Regional Alliance (TERRA) for Figure 9.1; and Subvertise for the use of Figure 6.2. Michael Roach kindly prepared Figure 9.2; and Larry McGrath provided Figures 9.3 and 9.4. Martin Wurt of Oxfam Australia was extremely helpful in locating illustrations, and must be thanked for providing and obtaining permission for the use of Figures 2.2, 3.2 (photographed by David Sproule), 5.1 (photographed by James Hawkins), 5.2, 6.1 and 7.2; the last consisted of photographs taken by Paul Weinberg. All have been generous in giving their time and permission to allow the use of these images. Every effort has been made to contact copyright holders to obtain their permission to reprint material in this book. The publishers would be grateful to hear from any copyright holder who is not acknowledged and will undertake to rectify any errors or omissions in future editions of this book.

Finally, the manuscript would not have been completed without the support and encouragement of family and friends, including Anya, whose enthusiasm for the project and tolerance of my long hours at the word processor facilitated the editing process, David, who assisted with some laborious routine editing tasks, and my unfailingly supportive parents, Denise and Douglas Percival. John Connell's ever-cheerful encouragement, insights and constructive comments on drafts contributed much to its final form. Colin Hearfield's meticulous editing and Deb Vale's assistance with compilation of parts of the draft were also invaluable. Zoe Kruze, formerly of Routledge, must be thanked for her prompt and enthusiastic advice early in the gestation of this volume, as should Jennifer Page, also of Routledge, for her patience, perseverance and assistance throughout most of the production process.

What is regrettably missing from this volume is a Southern perspective. The work primarily presents Northern views about the work of Northern organisations and their impacts, largely about the policies and practices of Northern institutions seeking to influence the 'development' process in Southern nations. In that sense, this volume presents an incomplete story of the growing influence of the impacts of the advocacy work of Northern NGOs. The authors nonetheless hope it will contribute to greater understanding of a relatively new, but increasingly powerful, global trend.

Barbara Rugendyke
December 2006

Acronyms and abbreviations

ABA	Asian Bureau Australia
ABC	Australian Broadcasting Corporation
ACC	Advocacy Coordinating Committee (of Oxfam International)
ACCORD	African Centre for the Constructive Resolution of Disputes
ACFID	Australian Council for International Development
ACFOA	Australian Council for Overseas Aid
ADAA	Australian Development Assistance Agency
ADAB	Australian Development Assistance Bureau
ADB	Asian Development Bank
AFP	Agence France-Presse
AIDAB	Australian International Development Assistance Bureau
AMRC	Asia Monitor Resource Centre
*Angli*CORD	Anglicans Cooperating in Overseas Relief and Development
AODA	Australian Official Development Assistance
APACE	Appropriate Technology and Community Awareness
APACE-VFEG	Appropriate Technology and Community Awareness – Village First Electrification Group
AREA	Association for Research and Environmental Aid
ARHA	Australian Reproductive Health Alliance
ASEAN	Association of South East Asian Nations
AusAID	Australian Agency for International Development
AVA	Australian Volunteers Abroad
AWD	Action for World Development
BBC	British Broadcasting Commission
BINGOs	big international non-government organisations
BOND	British Overseas NGOs for Development
BP	Beyond Petroleum (formerly British Petroleum)
CAA	Community Aid Abroad (now Oxfam Australia)
CAFOD	Catholic Action for Development
CARE	Co-operative for Assistance and Relief Everywhere
CDCAC	Canadian Democracy and Corporate Accountability Commission

CEO	chief executive officer
CSO	civil society organisation
CUSO	Canadian University Service Overseas (CUSO is now the stand-alone name for the organisation)
DFID	Department for International Development (UK)
ECOSOC	Economic and Social Council of the United Nations
ED	executive director
EdL	Électricité du Laos
EFA	Education for All
EMD	Environmental Management Division (of the Mitigation and Compensation Program of the Theun-Hinboun Hydropower Project)
ENGO	environmental non-government organisation
ESAF	Enhanced Structural Adjustment Facility
ESRC	Economic and Social Research Council (UK)
EU	European Union
EURODAD	European Network on Debt and Development
FACT	Fisheries Action Coalition Team
FDC	Foundation for Development Cooperation
FIELD	Foundation for International Environmental Law and Development
FIVAS	Association for International Water and Forest Studies
FOMACOP	Forestry Management and Conservation Project
G7	Group of Seven (group of seven industrially advanced nations: Canada, France, Germany, Italy, Japan, the United Kingdom and the United States of America)
G8	Group of Eight (an international forum for the governments of Canada, France, Germany, Italy, Japan, the United Kingdom, the United States of America and Russia)
GB	Great Britain
GCAP	Global Call to Action against Poverty
GCT	Global Coordination Team (of the Advocacy Coordinating Committee of Oxfam International)
GDP	gross domestic product
GM	General Motors
GMS	Greater Mekong Subregion
GNP	gross national product
GoL	Government of Laos
HDW	Human Development Window
HIDNA	HIV/AIDS Development Network Australia
HIPC	heavily indebted poor countries
HIV/AIDS	human immunodeficiency virus/acquired immunodeficiency syndrome
ICPD	United Nations Population Fund's International Conference on Population and Development

ICVA	International Council for Voluntary Action
IDA	International Development Association (of the World Bank)
IDEC	International Disaster Emergencies Committee
IDS	Institute of Development Studies
IMF	International Monetary Fund
INGOs	international non-government organisations
INTRAC	International NGO Training and Research Centre
IRN	International Rivers Network
IRRM	International Rural Reconstruction Movement
ISO	International Standards Organisation
ISSS	International Seminars Support Team
IWDA	International Women's Development Agency
KPMG	Klynveld Peat Marwick Goerdeler
MCP	mitigation and compensation programme (of the Theun-Hinboun Hydropower Project)
MDB	multilateral development bank
MDX Lao	a branch of a Thai infrastructure development company (formerly MDX Power and now GMS Power)
MIT	Massachusetts Institute of Technology
MLAs	multilateral agencies
MoU	Memorandum of Understanding
NGDO	non-government development organisation
NGO	non-government organisation
NOVIB	Netherlands Organisation for International Development Co-operation
NPV	net present value
OA	Oxfam America
OCAA	Oxfam Community Aid Abroad (formerly CAA, now Oxfam Australia)
ODA	overseas development aid
ODI	Overseas Development Institute
OECD	Organisation for Economic Co-operation and Development
OI	Oxfam International
OIED	executive director of Oxfam International
PDR	People's Democratic Republic
PNG	Papua New Guinea
PRGF	Poverty Reduction and Growth Facility
PRSP	Poverty Reduction Strategy Paper
PSS	Project Subsidy Scheme
PVC	polyvinyl chloride
TEAR	Transformation, Empowerment, Advocacy, Relief (Christian Aid organisation)
TERRA	Towards Ecological Recovery and Regional Alliance
TEW	Toward Ethnic Women
THPC	Theun-Hinboun Power Company

TNC	transnational corporation
TRRM	Thai Rural Reconstruction Movement
UNDP	United Nations Development Programme
UNECA	United Nations Economic Commission for Africa
UNEP	United Nations Environment Programme
UNICEF	United Nations International Children and Education Fund
VSO	Voluntary Service Overseas
VUSTA	Vietnam Union of Science and Technology Associations
WADNA	Women and Development Network of Australia
WAO	Washington Advocacy Office (of Oxfam International)
WCD	World Commission on Dams
WDM	World Development Movement
WSSD	World Summit on Sustainable Development
WVA	World Vision Australia
WWF	World Wildlife Fund for Nature

1 Lilliputians or leviathans?

NGOs as advocates

Barbara Rugendyke

> The Global Call to Action against Poverty can take its place as a public movement alongside the movement to abolish slavery and the international solidarity against apartheid. . . . Like slavery and apartheid, poverty is man-made and it can be overcome and eradicated by the actions of human beings.
>
> (Nelson Mandela, 3 February 2005, at the launch of the Make Poverty History campaign in Trafalgar Square, London)

What did my 15-year-old daughter have in common with Prime Minister Tony Blair, Nicole Kidman, Kate Moss, the rapper P. Diddy and Nelson Mandela in 2005? They all wore white 'Make Poverty History' wristbands, as did all of her schoolmates at the secondary school she attended in Oxford in that year. The wristbands were so 'cool' that peer pressure to have one was intense. Being publicly seen to be a supporter of a non-government organisation (NGO) or of a campaign supported by NGOs has become trendy in parts of the Western world, whether you are a prime minister, model, rock star or school student, or one of the world's greatest human rights activists. While wearing a bit of plastic may seem to be tokenism, the advocacy work of non-government organisations has become an increasingly important global phenomenon. So important is it that former US president Bill Clinton recently ranked the influence of NGOs, along with the extension of democracy and the internet, as one of the three global changes since the demise of the Cold World which give ordinary people the capacity to effect change in the world:

> There will always be problems in the world. . . . But because of the rise of non-government organisations in a world that is more democratic, in a world where the internet gives people more access to information, we don't have the excuse that we can't do anything about the problems we care about because the people we voted for in the last election didn't win.
>
> (Clinton 2006: 13)

Representing local, regional and national constituencies in Western nations, NGOs historically acted primarily at the local scale in 'developing' or Southern nations as they sought to improve the quality of life of people in disadvantaged communities. However, in little more than a decade, there has been a major shift in NGO practice; where once NGOs concentrated their work on establishing projects to do things like build water supplies or encourage income generation, the same NGOs have increasingly devoted resources to advocacy campaigns directed at global actors such as the World Bank, the International Monetary Fund, the World Trade Organization and multinational corporations. In so doing, and facilitated by advances in communications technologies, NGOs have themselves globalised, forming new strategic alliances in order to maximise their impact.

Simultaneously, the growing cooperation of NGOs in global advocacy campaigns has resulted in such groundbreaking and highly successful campaigns as Jubilee 2000, which mobilised 24 million people internationally under the slogan 'Drop the Debt'. When 24 million people from over 60 countries sign a petition, politicians take notice (Mayo 2005: 174). After people lobbied the G7 meeting in 1998, world leaders at the Cologne G8 summit in 1999 agreed to cancel $100 billion of debts owed by the poorest nations (Bedell 2005). The continuing Make Poverty History campaign is a UK-based national movement which is part of a worldwide movement – the Global Call to Action against Poverty (GCAP) – which not only targets those wielding political and financial might, but also strengthens NGOs to build broad-based public support for their causes. In 2005, the Make Poverty History campaign had 540 member organisations in the United Kingdom alone committed to it, including charities, development NGOs, trade unions and faith communities. Only six months after its launch, 87 per cent of the United Kingdom's population was aware of the campaign and 8 million people in the UK wore its white wristband (makepovertyhistory 2006). The Global Call to Action against Poverty coalition involves organisations in over 100 countries around the world and claims to have mobilised 53.5 million people to take action in support of its aims to tackle global poverty by lobbying for trade justice, more and better aid and further debt cancellation (GCAP 2006).

The effects of the Make Poverty History campaign were abundantly obvious in the United Kingdom during 2005. The Make Trade Fair component resulted in supermarkets vying to sell fair trade products, county councils producing brochures to tell the public where they could obtain fair trade produce, churches advertising themselves as 'fair trade' churches and displaying 'Make Poverty History' banners, and fair trade fashion parades and fair trade markets being held. In April 2005, 250,000 people took part in an overnight vigil for trade justice in Westminster. In Fair Trade Week, the public were barraged with publicity about the importance of trade justice, from human 'bananas' parading the streets, to street stalls and a concerted media campaign. Some muesli packets described where every ingredient was sourced, who had produced it, and how the purchase of the pack had directly assisted producers in Africa. Consumers could not only feel it doing them good, but feel they were doing the world good by eating the product! Sainsbury's supermarket chain in Britain reported a 70 per cent increase in fair trade sales over a 12-month period, fair trade turnover

Figure 1.1 Welsh campaigners on the Make Poverty History march, 2 July 2005, Edinburgh

has been growing by 20 per cent a year in Europe since 2000, and sales are projected to grow by 50 per cent per annum in Australia (Phillips 2006). Fair trade seems to be catching on, perhaps because responding to this focus of NGO advocacy gives ordinary people a sense of 'belonging', or of responsibility and direct involvement – their consumption, they hope, will result in direct benefits to poor producers.

That poverty in Africa was a major UK general election issue in 2005, to which the media gave much air time and column space, resulted from persistent pressure from the NGOs. That it was on the agenda at the Gleneagles G8 meeting was also largely a result of concerted lobbying by the NGO community, assisted by the associated Live 8 concert at Gleneagles (Gosch 2005): a 'bizarre confluence of politics and populism within the UK' (Geldof 2005: xxviii). An estimated 250,000 people took to the street in Edinburgh to demand world leaders take action to make poverty history (makepovertyhistory 2006). Although reactions by NGOs to the G8's decisions related to debt relief were mixed – 'the people have roared but the G8 has whispered' – that debt relief was so firmly on the agenda at all is widely attributed to the Make Poverty History campaign (Button 2005; Elliott 2005; Hodkinson 2005: 15). Thus NGO attempts to influence public opinion in order to influence national and global leaders to undertake policy changes are clearly having an impact and seem to be realising Clark's early speculation about the growing concentration by NGOs on advocacy: 'If it were possible to assess the value to the poor of all such reforms they might be worth more than all the financial contributions made by NGOs' (1991: 150).

Much has been written about the increasing emphasis placed on advocacy by NGOs and the ways in which communications technologies have enabled NGOs or other activist organisations to build new strategic alliances to assist them to advocate for change (Wiseberg 2001; Leipold 2002; Clark 2003; McCaughey and Ayers 2003; Meikle 2003; Rolfe 2005). Media reports about the effectiveness or otherwise of global campaigns abound (Button 2005; Hodkinson 2005; Nason and Lewis 2005). However, few detailed studies have been undertaken as a basis for gauging the extent to which NGOs 'going global' with their advocacy work has contributed to poverty alleviation: 'since advocacy . . . is relatively new and has evolved into a very dynamic process, there is not a lot of empirical data available on the extent of advocacy within NGOs' (Lindenberg and Bryant 2001: 205). Some authors have recently called for research to 'improve our understanding of NGOs as both subjects of development research and as actors in development processes, since these are inextricably linked' (Lewis and Opoku-Mensah 2006: 674). Some writings about NGOs have been justly criticised because 'More often than not they [NGOs] escape scrutiny and are simply posited as alternative signs of hope against dominant development discourse' or because comparative studies designed to reach a wide readership 'can only scratch the surface of each of the cases' (Hilhorst 2003: 2–3). To date, there is little empirical research to use as a basis for answering the question: just how effective is NGO advocacy in achieving changes which contribute to improved quality of life for poor communities in the world's poorer nations? Hence this volume. While it cannot address that larger question in its entirety, it brings together empirical work about the extension and the impacts of the advocacy efforts of development NGOs, which is largely previously unpublished and which 'better reflects empirical realities of the world of NGOs' (Lewis and Opoku-Mensah 2006: 673). This book presents evidence for the growing effectiveness of NGOs in harnessing greater public support for their goal of achieving greater equity, and in influencing the policies and practices of global institutions. It also describes the complexities of the resulting relationships. In doing so, it explains a vitally important global trend for in this NGO activity, according to the NGOs (and to world leaders like Nelson Mandela and Bill Clinton), lies the potential for civil society to impact on the global and national institutions and associated structures and systems which perpetuate poverty by determining access to resources and power.

Within this volume, the NGOs referred to are those which are not for profit, are usually based within 'Northern' nations and are not self-serving. Although many now receive some funding from government sources, their management is independent of government and those which are the focus of this book, as their primary mandate, seek 'to relieve suffering and promote development in poor areas, especially Southern countries' (Sogge 1996a: 3). Such NGOs not only transfer materials and resources to the South, but they also transfer information and, increasingly, engage in lobbying and campaigning work in pursuit of their broad objectives of poverty alleviation.

NGOs' influence enlarged

NGOs have increased in popularity as a means of seeking to improve quality of life for those for whom poverty and disadvantage are a daily reality. Their diversity, the formation of new organisations, and the demise or change in name or focus of others, along with their dispersed nature, make it extremely difficult to collect accurate data about their numbers and size. However, many commentators have recorded growing numbers of NGOs across the globe. During the 1980s, the number of NGOs registered in OECD countries increased from 1,700 in 1981 to 4,000 in 1988 (Porter 1990). Northern NGO spending increased from US$2.8 billion in 1980 to US$5.7 billion by 1993 (Edwards and Hulme 1995). Moreover, in 2001, the grants distributed by international and Northern NGOs from OECD member countries were estimated to be US$10.4 billion (OECD 2002). By 2000, at least 35,000 NGOs were believed to be working internationally (Edwards 2002). As they have grown in number, they have grown in size. Lewis and Opoku-Mensah (2006) reported that a *Newsweek* article in September 2005, based on data from a Johns Hopkins University study, emphasised that NGOs have become 'big business' and, by 2002 and across 37 nations, their total estimated operating expenditure was US$1.6 trillion.

Active NGOs have also proliferated throughout the developing world, with a large number of new organisations formed to work in service delivery and, increasingly, to engage in campaigning. In 1990, over 10,000 NGOs were registered as development organisations in Bangladesh alone (Williams 1990). An increase of 60 per cent in numbers of indigenous NGOs was reported in Botswana between 1985 and 1999 and there was a 115 per cent increase in local NGOs in Kenya between 1978 and 1987 (Fowler 1991). Similarly, the number of registered NGOs in Nepal increased from 220 in 1990 to 1,210 in 1993 and in Tunisia from 1,886 in 1988 to 5,186 in 1991 (Edwards 2004: 21). It has even been suggested that India has at least 2 million active NGOs (Wikipedia 2006a). It goes almost without saying that, with increasing numbers of NGOs operating throughout the globe and increased funds being distributed by them, NGOs have been increasingly influential. But why have NGOs increased so in popularity and influence?

NGOs have a number of claimed advantages: they are able to be more flexible and innovative and respond to need more quickly than bilateral and multilateral donors; they can implement and administer projects and programmes at lower cost than other aid delivery bodies; they are more likely to work with and through local institutions; they are more likely to emphasise processes of change and skills learnt rather than provision of quantifiable tangible goods (the preference of official donors); and they are more likely to take risks associated with working in geographically remote areas, sectors neglected by governments, or politically unpopular areas (Tendler 1982; Rugendyke 1994). Perhaps most important among the 'articles of faith' which rapidly became accepted about NGOs is that they claim to have better links with the neediest groups in poor communities and regions of the world. Not constrained by having to work through governments, they are able to work directly with the poor using participatory, 'bottom-up' processes of project identification and implementation, based on longer experience of working with local people and more accurate knowledge and understanding of local needs and capabilities. Critically, given the focus of this book, their independence from government allows them to engage in lobbying and campaigning in pursuit of greater global equity and social justice.

Shifting paradigms in development theory have also accorded NGOs new status. These changes have been detailed extensively elsewhere (Rugendyke 1994; Anderson 2003; Ollif 2003), so will only be given cursory attention here. Disenchantment with growth and modernisation theories of development which predominated during the 1960s and 1970s and with the later dominant radical development theories which contributed to new understandings that development, as both condition and process, was perpetuated by inequitable global structures meant that by the mid-1980s theorists were referring to an 'impasse' in development theory (Booth 1985; Sklair 1988; Corbridge 1990). During the 1980s, a strand of thought which had existed parallel to the large-scale deterministic theories and emphasised the ability of people to bring about change, through what was later called 'human agency', became increasingly influential (Giddens 1982; Watts 1988; Long 1992). Alongside abstract development theory, which largely described macro-level structural change, debate about the most appropriate forms of development practice did not subside. During the 1980s, commentary about development practice increasingly prioritised the involvement of local communities at every stage of development planning. With broad appeal across the political spectrum, and variously called 'empowerment', 'participatory development', 'democratisation' or 'populism', this emphasis recognised that social movements may be the fundamental agents of social change and opened the way for greater interest in non-state actors, including NGOs, in the development process (Corbridge 1992; Slater 1992). Much has since been written about the role of civil society in promoting positive change at every scale – from communities to nations and the global scene – and history has since demonstrated the enormous potential of 'people power' to effect change (Clark 2003; Keane 2003). This new focus on civil society, on the role of non-state actors in bringing about change, gave new legitimacy to the participatory approach of NGOs (Edwards and Gaventa 2001; Edwards 2004; Potter *et al.* 2004). More recently, their ascendancy and new popularity with governments and official aid agencies have also been seen to be directly related to their relevance to a larger neoliberal economic and political agenda (Pearce 2000; Craig and Porter 2006; Robbins 2006).

The persistence of this alternative view of development and the new status it accorded NGOs, along with recognition of their benefits and disillusionment with larger aid agencies, given the failures of many large-scale development projects, meant that from the 1980s NGOs were increasingly supported financially by both bilateral and multilateral donors. With growing numbers and income, NGOs became increasingly influential. However, that influence was accompanied by greater scrutiny of their activities, which questioned their impacts, legitimacy and transparency, and by stronger demands for accountability (Clark 1991; Edwards and Hulme 1992, 1995; Smillie and Helmich 1993; Smillie 1995; Sogge 1996a).

Recognising their increased importance globally, NGOs have engaged in an ongoing struggle to find the best ways to work towards global equity. Staff of development NGOs continue to grapple with the dilemmas and uncertainties of development 'on the ground' in disadvantaged communities. While that has often been seen to be both their focus and their strength, they have been criticised equally for having a range of weaknesses, one of which is that, historically, they have had little impact at a global level, with accusations made in the early 1990s that they had failed to address the wider-scale structural causes of poverty (Bebbington and Farrington 1992; Edwards and Hulme 1992). In the following decade, NGOs sought to remedy this by increasing their commitment to advocacy, convinced this was the way to maximise both the impacts and the cost-effectiveness of their work (Edwards 2002).

The term 'advocacy' is generally used by NGOs to refer to campaigning, which involves attempts to change public opinion, and lobbying, which aims to change 'structures, policies and practices which institutionalise poverty and related injustice' (Anderson 2003: 35). Campaigning encourages public support for lobbying activities, so, in essence, both attempt to influence policy formation as a means of facilitating positive change in people's lives. While there is a diversity of approaches to advocacy, it is 'self-evidently of a political nature (both in itself and in terms of what it seeks to achieve)' (Eade 2002: x).

Expanding horizons: globalising NGO advocacy

Reasons for the growth in advocacy activity have been well documented (Edwards and Hulme 2000; Pearce 2000; Chapman and Fisher 2002; Edwards 2002), but it primarily arose from 'the realisation that development and humanitarian relief projects will never, in and of themselves, bring about lasting changes in the structures which create and perpetuate poverty and injustice' (Eade 2002: ix). Thus, the increased commitment to advocacy occurred, in part, in response to organisational learning resulting from their own experiences in the field, but also as a result of debates among development theorists during the 1970s and 1980s about the causes of 'underdevelopment'. The political and economic causes of underdevelopment were identified as such things as unfair terms of trade, low commodity prices, oppressive debt burdens and the uneven distribution of land and other resources among different social groups. As it became increasingly obvious that it was impossible to address these causes simply by funding development projects in disadvantaged communities, NGOs began to realise that they also needed to tackle the wider processes which contribute to ongoing global inequity, to educate about these issues and to lobby for change at national and international levels. A shift began to occur from working to alleviate the symptoms of poverty to transforming the institutions and values that cause those symptoms; hence, NGOs strengthened their efforts to influence global systems and policy.

NGOs also concentrated more on advocacy because of calls to do so from their Southern partners. In the early 1980s, in response to a question about what the British could do to help poor people in Tanzania, Julius Nyerere responded 'Change public opinion in your own country' (in Burnell 1991: 240). Since the 1980s, Southern NGOs have consistently called for Northern NGOs to engage in campaigning and lobbying as the primary means of expressing concern about the global crises of poverty, environmental destruction and social disintegration. This was accompanied by growing recognition that Southern NGOs were best placed to engage in project work at the local level, so Northern NGOs looked for new ways to contribute to poverty alleviation (Chapman and Fisher 2002).

Illustrating the growth of importance of NGOs and of their increasing commitment to advocacy as a strategy for achieving their goals, Part I of this volume presents a comprehensive account of the Australian development NGO movement. While recent articles confirm that NGOs remain 'an important and large-scale presence on the landscape of international development', a key limitation of research about NGOs has been an 'over-emphasis on organisational case studies which are rich in detail, but lacking in contextualisation' (Lewis and Opoku-Mensah 2006: 665–6). This has led to calls for the abandonment of a

> . . . dominant ahistorical and non-contextual research tradition that has been heavily influenced by a mixture of normative-political agendas . . . always . . . concerned with how to improve the work of NGOs, and has often carried an analytical perspective that has been limited to the impacts of NGOs at project level, to the advocacy capabilities of individual NGOs, or to the capacity of NGOs to work for the poor.
>
> (Tvedt 2006: 690)

In addressing some of these concerns, Part I of this volume is neither ahistorical nor non-contextual. Based on extensive empirical research, in Chapter 2 Cathryn Ollif and I provide detailed illustration of the growth and development of the NGO movement in one Northern (albeit geographically southern!) donor nation, which is broadly representative of that which has occurred in other donor nations. Thus, the growth of Australian NGOs is traced through Korten's well-known generational changes, from the provision of direct relief or welfare to the poor, to strengthening the capacity of the poor for self-reliance, to the expansion of operations into addressing the structural causes of poverty through advocacy work, to increasing engagement in policy advocacy (Korten 1987, 1990). The historical account of these shifts in NGO priorities demonstrates that context is important, as the politicisation of Australian NGO activity was, in part, prompted by regional concerns. Otherwise similar trends have occurred elsewhere.

In more recent decades, broader processes of economic globalisation have been mirrored in the increasing globalisation of NGO activity, marked by the formation of global NGO alliances and increased engagement in transnational networks (Yanacopulos 2002; Roberts *et al.* 2005), exchange of information, and global advocacy campaigns, such as the highly successful Jubilee 2000 campaign and the current Global Call to Action against Poverty. In Chapter 3, Ollif presents a précis of the advocacy work of Australian NGOs, describing the ways in which a number of NGOs have sought to influence public opinion within Australia and have, more recently, 'globalised' by participating in global alliances and campaigns. A number of key themes which recur in literature about NGOs and advocacy, and in later chapters throughout this volume, emerge there, including the legitimacy of NGOs to act as advocates on behalf of the poor, the accountability of NGOs for their advocacy goals and strategies, and the effectiveness of NGO advocacy and the extent to which organisations evaluate (or are able to evaluate) the effects of their lobbying and campaigning (Edwards *et al.* 1999; Eade 2002: xi; Nelson 2002).

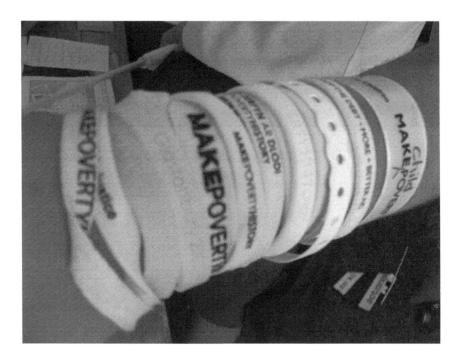

Figure 1.2 'Branding' anti-poverty campaigns: Make Poverty History wristbands

As NGO advocacy has continued to grow, there has been a corresponding increase in NGOs' influence in national and global affairs. Coordinating bodies have been formed nationally and internationally. Throughout the 1990s, Northern NGOs have put a growing proportion of their time and budget into advocacy, and many have created dedicated policy departments for research and lobbying. It is now standard practice for the World Bank and UN agencies to consult NGOs. In 1997, reflecting the growing influence of NGO voices, the UN Security Council held its first meeting with NGOs and, by 1998, the UN Secretary-General's report argued that 'NGOs are no longer seen only as disseminators of information, but as shapers of policy and indispensable bridges between the general public and the intergovernmental process' (UN report A/53/170, in Stephenson 2000: 291). At the United Nations, from 41 NGOs granted consultative status by the Economic and Social Council (ECOSOC) in 1948, and 377 in 1968, the number of NGOs in consultative status has expanded to over 1,550 (Opoku-Mensah 2001). Indeed, with the UN having 'realised the benefits of working with NGOs and sought to strengthen their relationships with them', NGOs have 'become increasingly *incorporated* into the UN system' (Martens 2006: 692).

Although NGOs are now widely acknowledged as being important in international affairs, indeed increasingly integrated into global decision-making processes, it is still 'extremely difficult to find comparative macro-level data on NGO types, activities and resource-flows [and] NGOs . . . are increasingly ubiquitous' (Lewis and Opoku-Mensah 2006: 666). Part II of this volume therefore broadens in focus, exploring the extent of the advocacy efforts of international NGOs and the contribution thus made to global policy formation. In Chapter 4, Ian Anderson presents comparative macro-level data from the period 1984 to 2005, collated to explore the extent to which NGOs have heeded calls to 'scale up' their advocacy (Edwards and Hulme 2000). To further illustrate the internationalisation of advocacy efforts, this chapter documents the merging of Oxfam affiliates into Oxfam International and the establishment of the Washington Advocacy Office of Oxfam International, formed to facilitate global coordination and expansion of the advocacy activities of the Oxfam affiliates.

Existing publications have widely acknowledged the policy reform successes of advocacy campaigns. These have tended to focus on the success of individual campaigns, such as the promotion of breastfeeding in Ghana and the campaign against child labour in the carpet industry in India (Chapman and Fisher 2002), or to list the achievements of national NGOs, such as claimed advocacy successes of UK NGOs in relation to marketing of baby milks, drafting of essential drug lists and the removal of restrictions on imports manufactured in the South. More generalised successes have included action related to global warming and rainforest destruction following NGO influence at major UN conferences (such at the 1992 Earth Summit and at the World Trade Organization meeting in Seattle), and general influences on the practices of multinational corporations related to employment conditions and to mitigation of and compensation for the social and environmental impacts of resources extraction or of large-scale infrastructure projects (Clark 1992; Anderson 2002; Edwards 2002). Much advocacy has been directed at the multilateral development banks, particularly related to debt relief and structural adjustment policies, and to reduction of the negative effects of large-scale infrastructure projects but, until recently, little was known about whether, through this, NGOs have had any impact in reducing the structural causes of poverty and were 'employing strategies which maximise their effectiveness and impact' (Anderson 2002). In Chapter 5, Anderson therefore explores the impacts of Oxfam advocacy on World Bank policies related to debt relief, and in attempting to urge a closer link between the World Bank's Heavily Indebted Poor Countries (HIPC) Initiative and poverty reduction strategy programmes in beneficiary countries. Anderson demonstrates that Oxfam's advocacy in relation to debt relief and Poverty Reduction Strategy Papers (PRSPs) had a material influence on relevant World Bank policy. From 1997, when Oxfam began its concerted campaign in relation to HIPC debt relief and poverty reduction programmes, a significant shift could be traced in World Bank policy towards Oxfam's proposals. Oxfam's influence was acknowledged not only by a range of stakeholders who were interviewed, but also in formal World Bank papers. Tracing the role of NGOs in processes which contributed to significant change in global policies is invaluable, and provides a more institutional perspective on the influence of Oxfam, as described recently by Mallaby in his account of changing World Bank practices under James Wolfensohn; there Oxfam was referred to, among others, as a 'grown up NGO' (2004: 55) and as 'the unofficial leader of the non-governmental aid groups' (2005: 60). Mallaby was unequivocal in his claims that NGOs have influenced World Bank policy and practice, although not always in positive ways: 'These constant NGO offensives tie up the World Bank, frequently disabling its efforts to fight poverty; despite their diminutive stature, the Lilliputians are winning' (2005: 6–7).

NGOs have also sought, in varied ways, to influence corporations active in developing nations. Concerned about the growing power and wealth of multi-nationals, many of which wield more financial might than national governments and have significant impacts on both people and their environments in developing nations, many NGOs have increased the extent of their engagement with corporations. There are typically three types of NGO/corporate interactions: NGOs as fundraisers, seeking corporate donations; NGOs as campaigners, critical of corporate practices; and, the newest relationship, that of NGOs working with companies in trying to improve the social impacts of their business practices, through assisting in the establishment of codes of conduct, fair trade practices and other sets of standards (Sayer 2003). NGOs largely have a common view of the role of corporations in the developing world, about their positive and negative influences on development and the ways in which they can improve their social and environmental performances. There is broad agreement about what corporate practices are incompatible with development and which business sectors are therefore unacceptable as donors, but there is great diversity in the nature of NGO relationships with companies. An emerging literature has traced these relatively new, but growing, interactions between NGOs and companies (Bendell 2000, 2005; Dhanarajan 2005; Frame 2005; Hayes and Walker 2005; Sayer 2005; Utting 2005; Eade and Sayer 2006). However, research about relationships between development NGOs and corporations is still relatively sparse. In Part III of this volume, in Chapter 6, John Sayer reviews the range of relationships emerging between NGOs and corporations as the former seek to encourage improved corporate behaviour which will impact positively on local economies and on the most disadvantaged groups within them.

NGOs are only now developing mechanisms to enable them to handle growing corporate engagement. They have developed strengths in recognising the impacts of companies and in vetting them as suitable for cooperation in development work, but frequently lack skills in realising the most effective ways of working with them, which inhibits practical cooperative work within developing nations. This lack is most acute where the NGO may have first contact with corporations. The great limitation on NGO engagement with companies is that NGO policies about relationships with the corporate sector often concentrate on risk reduction and protection of NGO reputations. The dilemma for NGOs is that working at the programme level with companies to change the way they work has the potential for direct, tangible and demonstrably beneficial change in the impacts made by large and influential corporations, yet to engage with companies inevitably brings risks for NGO reputations. However, to refuse to engage when a company accepts criticism and approaches the critic for assistance with improvement, and simply to continue the criticism, may also pose long-term reputational risks for the NGO. These issues are explored by Sayer in Chapter 7, where comparative assessment of the policies of a range of NGOs sheds light on the nature of their emerging relationships with corporations.

Advocacy is increasingly acknowledged as affecting the policies of governments and multilateral banks and there is increasing engagement between NGOs and corporations as the former pursue their broad objectives of improving economic and social conditions for disadvantaged communities. There has been a growing literature about civil society and state relationships in some Southern nations and regions (Maina 1998; Saravanamuttu 1998; Reimann 2002; Garbutt 2003; Chong *et al.* 2005) but NGO engagement in advocacy at the regional and national level in many Southern nations is still in its infancy. In Part IV, Philip Hirsch extends discussion beyond advocacy targeted at global development agencies and other powerful international actors to look specifically at NGO advocacy directed at developing country governments. In Chapter 8, his brief review of the international experience of local and international NGOs moving from community development into advocacy surrounding environmental and social justice issues is followed with case studies drawn from the nations of the Mekong Region. Following a particular focus on the emergence of Thailand's NGO sector and its connections with global advocacy coalitions, divisions within NGO movements in countries of the Mekong Region are illustrated, particularly within the local environmental movement. In this chapter, Hirsch explores the contingencies of advocacy in very different political systems, outlining the strategies of 'advocacy by any other means' within the more closed political systems of Vietnam, Laos and China, where conventional advocacy is proscribed. The chapter also discusses emerging advocacy within evolving civil society–state relations in Cambodia. Finally, the question of an embryonic regional Mekong civil society within this highly differentiated set of political-economic conditions is considered.

The advocacy activity of development NGOs is often regional in focus. Within the context set by Hirsch, in Chapter 9 Lindsay Soutar explores the dynamics of relationships between the Asian Development Bank (ADB) and NGOs in the Mekong Region. NGO strategies to reform the Asian Development Bank, Bank responses to NGO activity, and the changes in NGO/ADB relationships which have resulted are examined. The focus here is on a case study of the Asian Development Bank and of NGO encounters over the Bank-funded Theun-Hinboun Hydropower Project in Laos. Studies of anti-dam processes have explored the relationships between Northern and Southern NGOs (Rotham and Oliver 2002), the evolution of localised opposition which attracted a wide range of actors and organisations to become a regional political struggle (Hilhorst 2003) and the ability of NGOs to disrupt World Bank plans for dam development (Mallaby 2005). Here, Soutar traces the complexities of inter-relationships between NGOs, the Asian Development Bank, the Lao government and companies contracted to construct the Theun-Hinboun Dam.

What becomes explicit throughout this volume is that NGOs are a force no longer able to be ignored by governments, companies and international financial institutions. They continue to move from a primary position as agents of 'development as delivery' to being vehicles for international cooperation in the global arena, with advocacy increasingly significant as the mechanism for achieving this. Evidence about the successes of advocacy is still mixed (Nyamagasira 2002). Hence, in the following chapters the strengths, limitations and impacts of NGO advocacy at different scales (the local, regional and global) are discussed. Similarly, the interrelationships involved in operating as effective advocates for change at these different but, in a globalised world, inevitably intertwined scales of operation are reviewed.

Recent successes, such as the Make Poverty History campaign influencing the agenda at the G8 meeting in Gleneagles and the Global Call to Action against Poverty harnessing global public support in favour of debt cancellation and fair trade, mean that NGOs are likely to further heed calls to heighten their advocacy activity. The Make Poverty History advertisements on television confronted the viewer with the reality that one child dies every three seconds from poverty in our world. Every day, 50,000 people, primarily women and children, die from poverty-related causes. NGOs will continue to try to bring that daily global tragedy to the attention of the public and to increase the impacts of their advocacy activities on those in a position to bring about the changes necessary to improve the lives of those less fortunate. The following chapters trace the extent, complexities and efficacy of some of the advocacy they have undertaken thus far, either individually or in global alliances, and the steady movement from being local to becoming global actors, in both scale and operations. Whether NGOs will remain Lilliputians or whether, through the global alliances they have formed, they will become leviathans on the world stage remains to be seen.

Part I

Contesting global futures

From charity to challenge

2   Charity to advocacy

Changing agendas of Australian NGOs

Barbara Rugendyke and Cathryn Ollif

Before the Second World War, Australian voluntary development assistance was minimal. The emergence of the voluntary aid movement in Australia, from its genesis in pre-Second World War missionary and charity agencies to the formation of development assistance organisations, is traced here. The growth in NGO activity, shifting fads in NGO priorities, the reasons for these and, in particular, the increasing priority given to advocacy activities are also documented, providing a comprehensive illustration of the historical changes which have resulted in increased prioritisation of advocacy activities by Australian NGOs.

The birth of a voluntary aid movement

As was common to the history of voluntary aid movements in Western countries (Lissner 1977: 58ff.), Christian missions were the earliest Australian organisations to provide material aid to the then colonies, although evangelism was the primary purpose of these organisations. Examples include the Australian Board of Missions, which operated informally from 1850 and officially after 1872; Interserve Australia, originally established as the Zenana Bible and Medical Mission Ltd in 1904; and the Australia Churches of Christ Overseas Mission Board Inc., which was formed in 1901. Confronted with the overwhelming needs of host communities, most missionary agencies saw themselves as having a dual responsibility in the world 'comprising both evangelism and social action – a concept which is laid upon us by the model of our Saviour's mission in the world' (Stott 1975: 34). Therefore, practical provision of welfare in the form of material goods and health and education services occurred alongside sharing of the Christian gospel. Needs of recipient communities were defined by expatriate missionaries, and assistance generally provided as a result of decisions made by Australian-based agencies. Solutions to needs of indigenous peoples were seen to lie in the provision of education and Western technology (especially health technology). However, by the late twentieth century, many missionary agencies had adopted both the 'project approach' and the jargon widely used by development practitioners and, increasingly, definition and implementation of development projects occurred at the request of partner churches. For example, in its early days of operation, Interserve provided medical and education services for women and children in India. This work was gradually extended to include a wide range of community health and development projects throughout Asia and the Middle East, only provided at the invitation of nationals.

The late nineteenth century also witnessed the formation of charities and service clubs, some of which subsequently became involved in the delivery of development assistance. Thus, the National Council of Australia of the St Vincent de Paul Society was set up in 1895 and the Young Women's Christian Association of Australia was formed in 1860, the latter to assist in educating women and to encourage them to reach their full potential. After the turn of the century, the National Council of the Young Men's Christian Association of Australia was formed in 1901 and Lions Club International was formed in 1917. These agencies were initially chiefly concerned with charitable works or community service within Australia, but today their activities also include development assistance. Thus, the Australian Society of St Vincent de Paul, in addition to work at home, assisted branches in Asia and Oceania with monthly contributions to enable the financing of small self-help projects, working 'to relieve the needs of deprived members of the community, be these needs material, physical, psychological or social . . .' (ACFOA 1988: 27).

A third group of agencies was established in response to the plight of refugees and orphaned children and the specific needs of those whose lives were devastated by wars or disasters. The International Red Cross was first established in 1863 in Britain and sporadically provided relief supplies to those affected by wars and disasters. Nine days after the outbreak of the First World War, the Australian branch of the Red Cross was formed; its primary purpose was to provide services on battlefields and in regions recovering from the devastation of war (Henry 1970: 24–6; Donovan 1977). Similarly, first founded in the United Kingdom, the Save the Children Fund assisted children who were victims of the First World War, and Foster Parents Plan, an international organisation, was established in 1937, initially to assist children displaced by the Spanish Civil War (Molumphy 1984). The work of both organisations was extended dramatically from an initial emphasis on charitable provision for individual children to support for the social and economic development of whole communities, a change reflected in a description of the work of Foster Parents Plan:

> In the shambles of postwar Europe the Foster Child was immediately identifiable by his new coat or sturdy shoes. . . . The little Greek boy, once legless and sullen on a Piraeus dock, scoots around with new legs on a shiny bicycle in an appeal from the 1950s. A similar appeal now might show the 'before' child standing in a dusty lane in front of a dilapidated little house. The 'after' photographs might look much the same. Looking more closely, one might see a cement well in the background which was not there before, or, perhaps, a few ducks or chickens. The little house might now have a new roof – not intrinsically dramatic. . . .

What would not show in the 'after' photograph is the fact that the child does not have intestinal parasites, or pneumonia, that a community garden provides the vegetables to prevent anaemia, that the child now has access to a safer water supply.

(Molumphy 1984: 308)

A post-war organisational boom

The Second World War stimulated the emergence of new aid programmes. While relief and reconstruction of war-torn Europe were the focus of official programmes, the plight of displaced persons prompted the formation of new voluntary organisations (Lissner 1977: 58–67; Webb 1977a; Hunt 1986: 7–8; Reid 1986: 8). Examples included the United Nations Association of Australia, established in 1946, which conducted a major post-war appeal for children in 1948; the Federation of Australian Jewish Welfare Societies which, primarily concerned with welfare work on behalf of Jewish people, commenced activities in 1947; and the Australian Council of the World Council of Churches, established in 1948, which was, in its early years, primarily concerned with support of refugees and provision of relief supplies (Henry 1970: 14–23).

The Second World War not only spawned a host of relief agencies, but it was also '. . . the catalyst which radically altered the power balance between the forces of colonialism and independence' (Tiffen *et al.* 1977: 82). The struggles and poverty facing many of the newly independent nations were transmitted to the developed world through improved communications systems and increased travel. Early official aid transfers were generally motivated by political and strategic interest, in particular the desire to establish diplomatic relations with newly independent nations and to prevent the spread of communism amongst them (Hunt 1986: 7–8). Corresponding to a growth in aid disbursements in the 1950s and 1960s was the emergence of theories supportive of foreign aid; the basic assumptions of most were that economic growth equalled development, that aid programmes could assist by helping to remove obstacles to economic growth and by providing necessary injections of capital, and that rapid economic growth would lead to benefits which would eventually 'trickle down' to the poor. Optimism prevailed about the ability of transfers of Western capital and technology to promote development and alleviate poverty.

The combination of increased official involvement in the developing world and growing affluence in the developed nations, coupled with new community awareness of the needs of the developing world and an emerging theoretical framework providing justification for aid activity, was reflected in rapid numerical growth of Australian NGOs. For example, the Lutheran World Service was formed in 1950, the Food for Peace Campaign (subsequently to become Community Aid Abroad and now Oxfam Australia) was established in 1953, and 1959 saw the formation of Australian Baptist World Aid and the Quaker Service Australia. This organisational boom continued into the next decade with For Those Who Have Less formed in 1962, Australian Catholic

Relief (now Caritas Australia) in 1964, World Vision Australia in 1966, and the Australian Foundation for Peoples of the South Pacific in 1967.

United Nations initiatives such as the proclamation of the First Development Decade in 1960 provided the impetus for fundraising appeals, some of which eventually became permanent organisations. For example, the World Refugee Appeal of 1960 provided the impetus for an ongoing campaign on behalf of refugees, resulting in the formation of Austcare in 1967. In 1961, the United Nations mounted another appeal – the Freedom from Hunger campaign. This resulted in a continuing Australian campaign under that name, concerned mainly, in its early years of operation, with agricultural development and food production, which continued to operate as an independent organisation until its merger with Community Aid Abroad in 1992 (Ollif 2003: 203). In 1961, the Overseas Service Bureau was formed to encourage skilled persons, through the Australian Volunteers Abroad scheme, to serve overseas. The Bureau hoped that sharing of expertise would encourage host communities to become self-sufficient.

The emphasis of voluntary aid agencies began to change during the late 1950s and early 1960s. Through the experience of delivering development assistance, agency staff realised that handouts of material goods to provide relief for individuals were only palliative measures. Such experiences led to a new rhetoric amongst Australian NGOs. Increasingly, their concern was to encourage communities to help themselves, to provide training or services which would benefit the community over the long term. Reflecting such broadened concerns and insights, the Food for Peace Campaign changed its name to Community Aid Abroad in 1962. Similarly, by 1967 the concerns of the Australian Council of Churches had widened from its primary focus on refugees and relief and it introduced expenditure on a range of development projects, including the provision of health facilities and personnel (Henry 1970: 17–19). Concurrent with these changes was an emerging emphasis on the effectiveness of aid, which became '. . . a constant concern of organisations . . . [which were] particularly anxious to find ways and means of ensuring its effectiveness from the view point of the recipient country' (Anderson 1964: 139).

Emphasis on the effectiveness of aid meant that the relationship between donor organisations and recipient communities began to assume increasing importance:

> A sensitivity to the recipients' attitudes to aid, and to the desirability of making aid a two-way process, wherever possible, was emphasised. . . .
>
> Many organisations try to incorporate in their aid programmes the opportunity for co-operative effort between the donor and the recipient. For example, Community Aid Abroad sponsors cooperative development ventures in India, with Indians themselves providing the leadership and organisation, and voluntary aid bodies the capital; Church organisations contribute funds to aid the development of national churches in Asia and elsewhere, leaving the latter a completely free hand in their use; service

clubs channel funds and other aid through their opposite numbers in Asian countries who recommend which projects should be helped in the first place.

(Anderson 1964: 129)

The 1960s thus commenced a significant period of growth in numbers of Australian NGOs formed (Rugendyke 1994: 55–7; Smillie 1999a: 40; Ollif 2003: 92) (see Figure 2.1). However, the decade was significant not only for the organisational 'baby boom' which occurred, but for the change in orientation of Australian NGOs from the primacy of relief aid and a charity mentality to concern with promoting long-term benefits for recipient groups and the effectiveness of aid endeavours, matched with the increasing participation of recipient communities in the aid relationship. Advocacy or lobbying activities did not then feature among NGO activities. It was also a characteristic of the NGO community in the mid-1960s that 'well established Church organisations provided the solid core of Australia's voluntary aid effort'; not only were Church-based agencies significant numerically, but 'Their influence and activity spread beyond the confines of the organisations . . . as they sponsor certain other aid organisations and are often represented on the latter's committees' (Anderson 1964: 129).

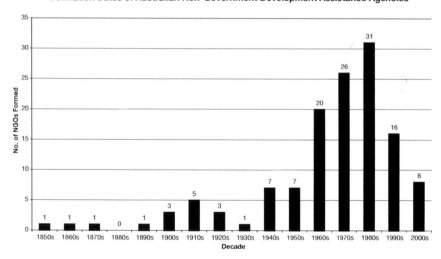

Formation Dates of Australian Non-Government Development Assistance Agencies

Figure 2.1 Formation dates of Australian non-government development assistance organisations

A new era: coordination and cooperation

Increased cooperation and communication (particularly in relation to joint fundraising campaigns) among Australian NGOs, and greater coordination of, and escalation in, lobbying and advocacy activities began in 1965 following the formation of the Australian Council for Overseas Aid (renamed the Australian Council for International Development, ACFID, in March 2004). In more recent times, the Council has acted as a vehicle for communication and negotiation with government, particularly in relation to government funding of NGO programmes and participation of NGOs in the official aid programme.

The origins of the Australian Council for Overseas Aid lay in a seminar of April 1964, held at the Australian National University in Canberra. Attended by representatives of 16 non-government aid agencies, the seminar was a first attempt to 'obtain an overall picture of the activities of the many and diverse Australian organisations engaged in foreign aid activities' (Anderson 1964: 127). One outcome of the meeting was recognition of the shared aims and values, and also problems, of the organisations and the possible benefits of cooperation and sharing of knowledge and experiences between them. This led to the conclusion that 'the time was ripe for the formation of a standing body in Australia to facilitate the exchange of information and ideas between voluntary organisations' (Anderson 1964: 141). Three months later, the voluntary organisations met again and appointed a 'Committee of Seven' to discuss the formation and constitution of an umbrella organisation for the NGO community. The Australian Council for Overseas Aid (ACFOA) was formally constituted in 1965. Reflecting strong community support for its work, the private aid community was growing rapidly, and it was in its interests to establish a forum for discussion, sharing of information, cooperation and coordination. The then Department of External Affairs also had an interest in the formation of a coordinating body for the NGO community. The Minister of External Affairs at the time, Sir Paul Hasluck, encouraged and facilitated the formation of ACFOA to create a vehicle to enable government to relate to the NGO community more easily (Henry 1970: 61, 77–8; Reid 1986: 90). Certainly, it did not take long for the Department of External Affairs to ask ACFOA to coordinate voluntary foreign aid operations in South Vietnam, providing finance for ACFOA to conduct a study of its aid needs (Henry 1970: 61).

The Council initially consisted of 20 member organisations, and was directed by a five-member committee. By 1989, ACFOA had 35 full member and 53 associate member organisations and by 2005 the membership of the Council, now ACFID, consisted of 80 organisations. The work of the Council was primarily sustained by annual subscriptions from members, grants from the commonwealth government, payment for services to non-members and donations (ACFID 2005: 15). ACFID policy is established at an annual council meeting consisting of representatives of each member organisation.

The very diversity of the NGO community constrained the Council's perfor-mance, and at times it was forced to adopt courses of action which failed to be approved by all members (Tiffen *et al.* 1977: 1–3). However, Council initiatives successfully stimulated considerable debate about development issues, and prompted change within the NGO community. Certainly, ACFOA became the major vehicle for communication between NGOs and government, and became a vehicle for coordination and cooperation amongst a growing community of very diverse members. The formation of the Council was thus a major event in the life of the evolving voluntary aid community, and its initiatives have had far-reaching implications for the NGO community as a whole.

The politicisation of Australian NGO activity

The 15 years after the formation of ACFOA were marked by increased politicisation of NGO activity. ACFOA activities stimulated debate which contributed to this process, as did the introduction of a separate official overseas aid programme (discussed hereafter). Of prime importance was a series of international events which demanded a response from Australian NGOs and prompted public controversy about their role.

Australian military involvement in the Vietnam War precipitated widespread debate and political activism within the Australian community, forcing questions about politics and aid on to the agenda of most Australian NGOs. The first government funds provided to ACFOA were for the coordination of voluntary aid activity to South Vietnam, which Australia had supported, along with the US, in the war against the communist North. While NGO efforts were largely concentrated in the South, when faced with the undeniable need of victims of war in North Vietnam many NGOs decided they could not ignore them and sent funds to the war-torn North. The ensuing controversy led to government attempts to prevent Australian aid being sent to that country. Accusations that major aid agencies had been infiltrated by communists were rife. Prominent personalities publicly affirmed their intention to donate funds to the North Vietnamese, while it was announced in Parliament that both the World Council of Churches and Australian Catholic Relief had sent money to North Vietnam (Henry 1970: 101). The controversy prompted many agencies to affirm a position that aid should be available to all in need, irrespective of the political stance of their governments; not to provide such aid was as political an act as to do so.

The debate about Australian involvement in Vietnam had even more profound implications for the NGO community. It contributed to a questioning of the involvement of developed nations in the developing world, and to a realisation that NGO activity at home, not only in the field, was required:

> Through the activities of anti-war activists many people have come to realize that . . . the rich become involved in military operations to suppress those elements in the Third World which would expropriate foreign companies and impose more stringent controls on the exploitation and use of their national resources. This understanding of the Vietnam War leads

people to see that 'action for development' starts in the rich world. We must reassess our own society and its relations with the Third World. . . . by looking at all aspects of Australia's relations with the Third World – military, diplomatic, trade, investment and aid.

(Newell 1972: 4)

Controversy erupted again over the involvement of voluntary agencies in campaigns to combat human rights abuses in the early 1970s. A tour of Australia by the South African Springbok rugby team in 1971 prompted a series of public demonstrations and widespread debate within the community about racism. The NGO movement became involved in campaigns against apartheid, and Australian Catholic Relief and the Australian Council of Churches supported the World Council of Churches in its Program to Combat Racism. The World Council of Churches donated funds to organisations fighting racism, to 'medical, education and social service activities of liberation movements in Angola, Mozambique, South Africa and Guinea-Bissau' (ACFOA 1973: 11). At the same time, member churches and Christian agencies were urged to press for a withdrawal of investment in, and cessation of trading with, South Africa. Support for liberation movements provoked outbursts from those who felt the World Council of Churches should not take a 'political stand' and to accusations that church agencies were supporting terrorists.

ACFOA publications throughout the 1970s featured reports on issues of major concern to its members – apartheid, racism and support of liberation movements. Many agencies became involved in campaigns to lobby the government, arguing that growth in Australian investments in South Africa forced the Australian business world and government into 'the position of upholding the present Government in South Africa . . . to protect their economic interests. The Australian Government therefore is positively supporting the inhuman system of apartheid as enforced by the South African Government' (Noone 1973: 6). ACFOA urged its members to take a stand and offer positive support to black South Africans in their struggles.

The activities of some Australian aid agencies in response to civil war in East Timor further fuelled controversy over the policy and actions of the Australian government. Concerns of Australian NGOs about Portuguese colonialism in the area were recorded from as early as 1968. Some agencies began to support self-determination for the East Timorese 'even before any of them had met any Timorese and before any political parties had been formed in East Timor' (Hill 1980a: 9). Shortly after a coup in East Timor in August 1975, the ACFOA annual Council meeting called on the Australian government to express support for the principle of self-determination for the East Timorese. On 28 November 1975, the East Timorese independence movement, Fretilin, declared East Timor an independent nation. However, this was not recognised by Portugal, Indonesia or Australia. A week after the full-scale invasion of East Timor by Indonesian troops on 7 December 1975, the ACFOA executive decided to take a stand strongly critical of Australian government policy; the latter considered integration with Indonesia to be in the best interests of East Timor (Hill 1980a: 8).

The resolute campaign by ACFOA against Australian recognition of Indonesian sovereignty over East Timor meant that Australian NGOs were not permitted to work in East Timor by the Indonesian government. The International Red Cross and US Catholic Relief Services were, in late 1979, allowed to begin work in East Timor. However, it was argued that both were 'conducting their programmes there on objectionable Indonesian Government terms' (Walsh 1980: 21). ACFOA and the majority of Australian NGOs were determined that their first priority was to fight for the rights of the East Timorese and to 'question the propriety of giving aid without justice' (Walsh 1980: 21).

The conflict in East Timor and the NGO response were significant for several reasons. This was the first occasion on which ACFOA called for the suspension of military aid to any country. In addition, ACFOA reached new audiences and the

> death of six [Australian] journalists and the fact that journalists as well as aid agencies are excluded from East Timor has given some journalists a new sense of identity with the aid agencies and more working relationships have evolved in relation to other parts of the world as well.
>
> (Hill 1980a: 14)

In an interview in 1990, a former Director of the Australian Council for Overseas Aid commented on the importance of East Timor as being:

> the first time aid agencies responded immediately and automatically to need and there was tremendous solidarity. . . . ACFOA and most member agencies made the decision that protecting the rights of the East Timorese was as important as the provision of relief.

Significantly, the incident provoked a sharp division between ACFOA and government policy, giving further impetus to the role of the voluntary aid community as a critic of government on moral and humanitarian grounds. Thus, throughout the 1970s, Australian NGOs increasingly engaged in lobbying and campaigning activities in response to significant regional and international events.

The emergence of dualism in NGO functions

While Australian involvement in international affairs precipitated considerable debate and change amongst Australian NGOs, the voluntary aid community was also influenced by a growing understanding of the processes which accounted for the persistence of poverty in the developing world. Change from a charity approach to an emphasis on needs based, self-help projects had occurred in their rhetoric as a result of the NGOs' own experiences in the field.

The late 1960s and early 1970s ushered in a period of widespread disillusionment with dominant theories of development and with the ability of aid to promote change, for the aid agency which tries

> to ensure that aid promotes 'development' is like a blind-folded man with his leg chained to the door post and one hand behind his back, in a totally unfamiliar room in which someone whom he cannot see and does not know keeps on moving the furniture around.
>
> (White 1971: 6)

Some development theorists rejected the notion that injections of capital could stimulate economic growth, stressing that entrenched inegalitarian structures could hinder development and that economic growth under such circumstances could increase inequalities in developing nations. The assumption that benefits of economic growth would necessarily trickle down to the poorest groups within developing countries was increasingly questioned, and this led '. . . to distribution questions being added to the agenda for development policy, and to the drawing up of needs based and welfare-specific targets'. Elsewhere, 'disillusion with the assumed method of alleviating poverty indirectly through accelerating the growth rate . . . led to an increasing emphasis on direct action through targeting aid to the poor' (Riddell 1987: 93).

Theorists argued increasingly that structural imbalances, such as an imbalance in international trade which favoured affluent nations, were the real factors hindering development, and that the distribution of aid could do little to effect the radical changes in international structures which could assist the poor nations. By the late 1970s, 'the central element of development studies was the view that underdevelopment or dependency, however defined, were not simply static conditions or symptoms, but products of dynamic processes, primarily engendered within the rich world' (Connell 1988: 1). Growing understanding of the interrelatedness of the rich and poor nations led to questioning of the relevance of aid projects. Thus, 'the rise of dependency and other critical-left approaches . . . reconceptualised the global relationship between First World and Third as exploitative rather than supportive. In this schema, the political development project ("trying to help") simply looked hypocritical' (Goldsworthy 1988: 55).

These debates contributed to recognition that aid had not begun to solve problems of world poverty and social justice, leading Australian NGOs to a fundamental re-examination of their role (Webb 1977a: 7). Reflecting broader discussions about global inequality, the debate within voluntary agencies increasingly became one focusing on trade and the workings of the international economic system. The task of NGOs was seen as extending beyond the disbursement of aid, to include the promotion of awareness and actions which could lead to significant changes in the structure of world society. Thus, an emphasis on action at home, not only on the distribution of aid, grew out of this shift in understanding, for:

Earlier, for most supporters of aid, the problem was seen to be 'out there' and the task was to transfer resources overseas as aid. Increasingly, however, the problem has been seen also to be 'over here', that is also within attitudes and structures of the donors.

(Webb 1977b: 1)

This changing orientation was expressed in a number of ways. Firstly, a number of new agencies were formed; their *raison d'être* was not to operate as funding organisations, but to initiate discussion and promote education of the Australian populace about issues of development and justice. For example, 1971 saw the genesis of Action for World Development (AWD), International Development Action and the Asian Bureau Australia (ABA). International Development Action sought to encourage Australians to prevent the Australian government from acting in developing countries 'in the defence of Australian interests against indigenous aspirations' (Newell 1972: 4). The ABA, a non-denominational Christian organisation, sought to educate Australians about the needs of those in Asia and the Pacific through engaging in 'research, documentation, public education and policy intervention, acting on matters of justice in the relationship between Australian people and our Asian and Pacific neighbours' (ACFOA 1988: 11).

Of particular significance in the early 1970s was the Action for World Development Campaign. The ecumenical agency was formed in 1971, as a joint venture of the Australian Council of Churches and the Roman Catholic Church, to organise a national education campaign, with the involvement of Australian church groups of all denominations. The aims of the campaign were to help Australians 'to a true appreciation of the aspirations and needs of other nations, of the ways by which we can hinder or help the development of others and of the actions which are possible for us as Australians' (ACFOA 1972: 3). Over 200,000 Australians participated in the national education campaign. Many current supporters of the voluntary aid movement regard the campaign as having had a major impact in confronting the public with issues of world development and in mobilising ongoing active support for the voluntary aid effort.

In the late 1970s, several more agencies sharing a primary concern with educating the public were formed. For example, the Development Education Group of South Australia was initiated in 1978 to coordinate and foster 'a process by which people are assisted to develop a critical awareness of the social, economic and political structures that affect . . . daily lives' (ACFOA 1988: 46); this involved publishing educational materials for use in schools and community groups, public seminars and workshops, and seeking to influence curricula for schools and adult education programmes. Trading Partners and the World Development Tea Co-operative, formed in 1978 and 1979 respectively, were also established with a mandate to educate the public, particularly about the injustice of international trading relationships. The World Development Tea Co-operative marketed tea in Australia as a means of building awareness of social justice issues and consumer resistance to the exploitative actions of

Figure 2.2 Reforming government aid: a focus of Australian NGO advocacy in the 1980s

Western multinationals operational in the developing world (Whelan 1982). Trading Partners, also an alternative marketing organisation, supported the self-help efforts of indigenous peoples by importing and selling their handicrafts and 'creating more awareness of peoples' struggles for better lives in the developing world, and linking this to the need for trade justice' (ACFOA 1988: 55).

The new emphasis on development education had far-reaching implications for existing voluntary aid agencies. In the aftermath of the Action for World Development Campaign, ACFOA's Education Unit (which had been established in early 1972) organised the first 'National Conference on Development Education'. Held in Canberra in January 1973, the conference attracted participants from educational institutions, aid agencies and lobby groups and included representatives from Papua New Guinea, Fiji and South-East Asia and from the Australian Aboriginal community. One observer suggested the conference was 'an extremely catalytic event. It ushered in a five year period of intense activity and high optimism in the field of development education' (Hill 1980b: 20). Following the impact of the AWD campaign and the education conference, and reinforced by a growing understanding of the factors hindering development, many agencies began to see the work of forming public opinion within Australia as of equal importance to funding overseas project work. Agencies employed development education officers who prepared materials for use in schools, churches and community groups; resource centres were established; educational films were produced; newsletters and magazines were published; agencies tried to influence the content of curricula for schools and tertiary institutions; and many urged their members to be more actively involved in lobbying activities.

The flurry of development education activities slowed in 1978, following an ACFOA summer school of development which was held in Tasmania in January of that year. A 'considerable uneasiness' observed at the conference 'made many participants among the NGOs question the value of all the centralization and co-ordination which ACFOA was doing' (Hill 1980b: 21). Waning enthusiasm for development education followed the ACFOA summer school of 1978, for

> . . . a widening gap had appeared between those with an established educational programme and rationale, and those who saw in the increasingly political content of those programmes, including that of the ACFOA education unit, some threat both to the way in which their aid programmes operate and to the whole rationale of aid, including the likelihood of withdrawal of public support and government grants because of the critical content of development action resulting from the issues raised through education.
>
> (Burns 1981: 36)

Despite lessening enthusiasm for the extension of development education activities and associated development action, there remained 'a basic dualism in the role of a voluntary aid agency in the 1980s . . . manifested in different ways throughout the range of an NGO's external and internal relationships' (Alliband 1983: 54). So, for example, the Australian Freedom from Hunger Campaign 'adhered to the former view of poverty being caused by endogenous factors, but over the past decade the second view of poverty being caused primarily by exploitative power relationships has gained predominance' and 'the conflicts between these views provide one of the principal internal dynamics of the organisation' (Alliband 1983: 54–5). Thus, many NGOs had two roles: a development assistance programme and a community education programme.

The emergence of new development paradigms and changes in development thinking by NGOs also prompted an examination of the effectiveness of aid programmes and a reappraisal of aid strategies. A number of principles became enshrined in the aid rhetoric, based on '. . . a modification of the growth objective to include equity and distributional concerns'. They included appropriate technology, employment generation, integrated rural development, women and development, participation and basic needs (Hunt 1986: 25).

Growing concern with the suitability of technology for local conditions and cultures was reflected in the formation of two Australian organisations, Appropriate Technology and Community Awareness (APACE, now APACE-Village First Electrification Group (VFEG)), formed in 1976, and the Association for Research and Environmental Aid (AREA), established in 1977. For both groups, capital-intensive Western technology was often inappropriate for communities in the developing world and they endeavoured to implement appropriate technology projects abroad. Today, APACE-VFEG continues to assist in 'ecologically sustainable development' (APACE-VFEG 2006). By the late 1970s, most Australian NGOs had rejected the assumption that Western technology was automatically the answer to problems of underdevelopment.

The shift from the delivery of relief supplies and Western technology to the search for long-term solutions for impoverished communities resulted in increased emphasis being placed on support for projects aiming to create employment, which often involved training, establishment of cottage industries in local communities and attempts to improve agricultural productivity through introducing new, appropriate forms of agriculture. The encouragement of self-sufficiency was central and the satisfaction of basic needs became the catch-cry of agency rhetoric (Hunt 1986: 21ff.). Through their field experiences, agencies learnt that involvement of local groups in processes of selecting and managing aid projects meant projects were more likely to be supported and successful. So participation of local communities in defining and implementing projects was increasingly stressed, to ensure local needs were met, rather than solutions imposed by outsiders.

Part of this participatory approach, now enshrined in the objectives of most voluntary aid agencies, was to ensure that the needs of women were not neglected in development planning. Two agencies with a specific mandate to ensure the participation of women in development efforts were formed in Australia during the United Nations Decade for Women (1976–1985). The Women and Development Network of Australia (WADNA) was formed in 1981, and the International Women's Development Agency (IWDA) in 1985, which today '. . . undertakes projects in partnership with women around the world . . . who suffer poverty and oppression' (IWDA 2006).

While it is beyond the scope of this brief historical account to explore changes in project orientation in depth, 'self-reliance, local control, partnership, appropriate technology, peoples' participation, including the participation of women, were all values which came very much to the fore as opposed to the "hand-out mentality", "band-aid type solutions", and huge capital intensive projects which engendered dependency' (Hill 1980b: 21). Major changes in agency orientation occurred throughout the 1970s:

> Since the beginning of the seventies there has been a radical shift in the approach to voluntary aid . . . from the approach typified by the 'bowl appeal' to actions designed to support movements for social changes which might remove the sources of inequality, poverty and oppression. This change has found expression in two ways. Firstly there has been increased emphasis in the rich countries on the need for 'structural aid' . . . establishment and support for rural, co-operative, self-help projects, support for liberation movements and the development of literacy programmes. Secondly, there has been an attempt to raise the consciousness and heighten the understanding of those in the 'rich' countries that impoverishment was the other side of their own enrichment: the root cause was to be found in the process of unequal exchange generated by the world capitalist system.
>
> (Sharp 1978: 47)

NGO/government cooperation

In the Australian context, the growth in the funding relationship of NGOs with the official government aid body, now known as the Australian Agency for International Development (AusAID), was of great importance. Australian government 'assistance' abroad was first evident in the transfer of resources to Papua New Guinea (PNG). Official links with the country, established in the 1880s, contributed to a concentration of Australian resources there, initially in administration during the pre-Second World War period, and later as grants of aid to the territory. Earliest aid allocations outside PNG were for post-war relief and rehabilitation programmes established by the United Nations. Subsequently, a major vehicle for disbursement of Australian Official Development Assistance (AODA) was the Colombo Plan, created in 1950 to promote cooperative support by Commonwealth nations for economic and social development in South and South-East Asian nations. Over the following two decades, aid allocations gradually expanded to include participation in international and regional bilateral aid programmes and increased support for multilateral operations. Until 1973, aid functions and staff were scattered through a number of federal and also state departments.

In September 1973, the Australian government established a separate statutory body to have responsibility for the administration of all Australian bilateral and multilateral aid. The new body, the Australian Development Assistance Agency (ADAA), was responsible to the Minister for Foreign Affairs, and absorbed the aid functions then carried out by the Department of Education, the Treasury, and the Department of External Affairs. ADAA, however, was short-lived. Following the dismissal of the Labor Government in November 1975, a policy of economic stringency led to the abolition of the Agency. The centralised administration of aid was maintained within a new organisation, the Australian Development Assistance Bureau (ADAB). The new organisation was part of the Department of Foreign Affairs, rather than an independent aid agency. ADAB administered the provision of aid to developing countries, formulated aid policy, and was granted a substantial degree of autonomy in relation to the financial management of overseas development aid (ODA). In 1987, ADAB was renamed the Australian International Development Assistance Bureau (AIDAB), and in 1995 it was given its current name, the Australian Agency for International Development (AusAID).

Government financial assistance was first given to the NGO community in 1965/66 in the form of grants to the Australian Volunteers Abroad (AVA) programme and the Australian Council for Overseas Aid (ACFOA). In 1974/75 a programme of regular financial assistance to Australian NGOs, which became known as the Project Subsidy Scheme, was introduced (ADAA 1975: 39). By 1983/84, official support for NGO activities, including the delivery of emergency aid and relief assistance, totalled A$10 million. By 1989/90, A$50 million were channelled through Australian NGOs (AIDAB 1990: 2) and by 2003/2004 total AusAID funding to Australian NGOs was nearly A$95 million (AusAID 2005a). Although only a small proportion of ODA is disbursed through NGOs, that funding is significant to many NGOs. Former ACFOA staff reported that some agencies historically derived as much as 70 per cent of their income from government subsidies.

Over time, various mechanisms in addition to the Project Subsidy Scheme provided funding through NGOs. The introduction of programme funding in 1990/91, through which eligible NGOs with a proven track record could receive block grants to spend on their own programmes, signified a new level of trust in the relationship between AIDAB and Australian NGOs, as well as reflecting the fact that programme funding was more easily administered than the selection and funding of individual projects. Funding for emergency relief and refugees accounted for a large proportion of total financial assistance given to Australian NGOs, which cooperated extensively with AIDAB (later AusAID) in the delivery of emergency relief and food aid. This included during crises in Kampuchea, Ethiopia, Mozambique and East Timor and, more recently, in the supply of over $12 million through NGOs to tsunami-ravaged South and South-East Asia in 2005 (AusAID 2005b: 96).

Government financial assistance has also been given to the NGO community indirectly. Before 1980, the federal government on occasion allowed tax deductibility for donations made to selected agencies or for specific appeals. In August 1980, it was announced that contributors to approved Australian NGOs would be able to claim donations above A$2 as an income tax deduction. By foregoing taxation revenue, the government encouraged the public to support the work of Australian NGOs, enabling them 'to significantly increase the practical and effective aid they are presently providing' (Peacock and Howard, cited in ADAB 1980: 86).

In addition to these forms of financial support of NGO activity, AIDAB and AusAID have encouraged NGOs to participate in its general development programme, including through its Women in Development Fund, funding for seminars relevant to development issues under the International Seminars Support Scheme (ISSS) and funding for community-based primary health care projects which focus on the health of women and children; and the NGO Environment Initiative was established to provide funding for environmental projects initiated by Australian NGOs (AIDAB 1990: 4). By 1980, in recognition of the quality and benefits of much NGO aid, ADAB began to explore ways of involving Australian NGOs in its bilateral programme, for 'the ability of NGOs to service some of the areas of most serious poverty and to operate at the grassroots level at which the official programme cannot readily function has long been recognised' (Peacock 1980: 9). Thus, from the 1980s, NGOs participated in the bilateral programme in various ways.

The level of government funding of voluntary agency activities thus increased steadily, as did the number of options available for NGO involvement in the official aid programme. By 1988, more than 20 per cent of Australian NGO funds were derived from government, in an increasingly complex relationship between government and the NGO community. For some NGO personnel it was imperative that NGOs espousing the benefits of a poverty-focused aid programme and the advantages of their aid should be willing to participate in the government aid programme to assist in the growth of an official programme with a clear poverty focus. Most agencies insisted that the availability of government funding was advantageous in allowing them to extend their overseas activities, and many argued that an ongoing relationship with AIDAB encouraged 'increased professionalism' amongst the voluntary organisations. Conversely, others argued that increased government financing of NGO programmes had not necessarily been positive and that rapidly increasing size and quantity of NGO projects resulted in a decrease in their quality (Bysouth 1986: 215; Nichols 1987).

Many were also concerned that acceptance of government funding could undermine the willingness of Australian NGOs to engage in lobbying or campaigning which might be critical of the Australian government. Thus, by the end of the 1980s, in the words of a former ACFOA chairperson:

> . . . the most crucial issue currently facing the voluntary aid community is 'The Management of the NGO Relationship with Government'. Putting it bluntly, the potential for government to effect changes in NGO perspectives

and programs in even the most benign relationship cannot be overstated.
. . . How we manage our relationship with government and how we translate
our idealism into practical strategies will show either the hollowness of our
words or the genuineness of our commitment to our partners in the Third
World.

(Ross 1988: 2–3)

The 1980s

The priorities and rhetoric of Australian NGOs were well established by the
1980s. Participation, partnership, a poverty focus, grassroots involvement,
provision of basic needs, appropriateness, innovativeness and support for social
justice remained central tenets of the voluntary aid movement. Significant events
in that decade did not change the primacy of these objectives, but expanded
public exposure to the voluntary aid community and widened the scope of NGO
activities.

From the perspective of the Australian public, probably the most significant
events involving Australian NGOs in the 1980s were major emergency relief
appeals. In the face of two crises, the NGOs pooled their resources and
approached the public with appeals for funds through the International Disaster
Emergencies Committee (IDEC) – formed in 1973 by ACFOA to coordinate
requests for public funds following a disaster or emergency in another nation.

Following graphic reports of suffering and starvation in Kampuchea
(Cambodia) by journalist John Pilger, IDEC announced the commencement
of the Kampuchean Relief Appeal in September 1979. Generating response
unprecedented by any appeal of its type in Australia, it raised A$10.5 million
from 3 million Australians in 1979/80. The close proximity of the country,
Australian involvement in the war in Vietnam, and horror at the treatment of
Khmers by the Pol Pot regime combined to give Australians a special commitment to the needs of the nation. Significantly, the income from the 1979
appeal provided agencies with extremely large amounts of money for the
first time, leading to expansion of their capacity and influence. The huge
public response and the urgency of the need in Kampuchea also convinced
the government to allow tax deductibility for donations – a first step towards the
extension of tax deductibility status to other voluntary agency activities. The
appeal resulted in a continuing and unique involvement of Australian voluntary
agencies in Kampuchea – the NGOs collectively accepted the international
responsibility for relief and then rehabilitation and redevelopment of the nation.
Australian agencies became involved in projects which were on a scale 'larger
than normal and which contributed to getting the system working again in
Cambodia. Irrigation canals, pumps, fertilizer plants, pharmaceutical plants,
power stations and many other areas were on the long list of needs' (Ashton
1989: 11). NGOs were able to undertake planning for large, long-term projects
with some certainty about funding, and a new level of coordination and
cooperation amongst Australian agencies resulted. In 1986, a joint Australian

NGO office was opened in Kampuchea, the first permanent Australian presence in the country since the crisis of 1979.

The activities of voluntary agencies from developed nations were thrust further into the public eye following massive media exposure of the extent of despair in the famine-devastated Horn of Africa. Many Australian agencies were already involved in Africa, but the need was so great that, in October 1984, IDEC launched the Ethiopian Famine Appeal. In six weeks, A$3.3 million were raised. In July 1985, following international Live Aid concerts, Australian agencies and IDEC became involved in the 'Oz for Africa Live Aid Telethon'. The event was significant because of its massive impact on the public and because of the unprecedented level of cooperation between NGOs, media and private industry (ACFOA 1985: 17). The tremendous public response elicited by the appeals had continued effects, prompting increased financial commitment to NGOs which lasted well beyond the media campaigns. In addition, the Australian government channelled A$7 million of emergency humanitarian and food aid through Australian voluntary agencies during 1984/85. Thus, '. . . the Government affirmed the capacity and effectiveness of NGOs in channelling food aid and emergency relief' (ACFOA 1985: 31).

The 1980s were also marked by a growing focus on the quality of aid by the NGO community. Increased involvement with ADAB gave voluntary agencies new insights into the official aid programme, prompting a spate of discussions on the quality of aid. As John Birch (then Chairman of ACFOA) wrote, 'Whilst the closer relationship between Government and the voluntary sector creates concern about lost autonomy, it also provides increased opportunities for effecting [*sic*] the direction and quality of the Government aid program' (Birch, in ACFOA 1985: 2). NGO commitment to constant lobbying about the level of government official aid was thus extended to include concern with the effectiveness of official aid in helping the poor.

In the early 1980s, evaluation reports examined the impacts of large government aid projects in the Philippines and found that large, integrated rural development projects benefited wealthier members of the communities, or the central government and its international partners, at the expense of the poor, reinforcing unjust social conditions (Richards 1981; Shoesmith 1982). Such findings catalysed debate within Australia about the quality of official aid. Concern with the quality of Australian ODA was also reflected in submissions made by voluntary agencies in 1984 to the Committee to Review Australia's Overseas Aid Program; the submissions criticised assumptions about development and aid inherent in the committee's report (the Jackson Report). The Jackson Report's recommendations 'for the doubling of funds allocated to small scale, community based projects undertaken by NGOs, as well as supporting the allocation of public funds for development education', did, however, recognise the quality of NGO activities (ACFOA 1985: 4). Outstanding in the memories of ACFOA and NGO staff active at the time is the commonality of purpose displayed by the voluntary aid community in submissions and reactions to the Jackson Report (ACFOA 1985: 2; Eldridge 1985; Forbes 1985; Jackson 1985; Stent 1985; Vale 1985; Bysouth 1986).

The concern of NGOs with the quality of aid was not limited to criticism of the official aid programme. Increased public exposure during the major disaster appeals in the first half of the decade, combined with growing involvement in the official aid programme, brought demands for greater accountability. Debate about the quality of voluntary aid focused on the importance of the evaluation of completed projects, as well as on the monitoring of ongoing projects, and explored the possibilities of participatory-style evaluation reflective of NGO philosophies (Porter and Clark 1985: 25). The development of more appropriate evaluative methods is a continuing process for Australian NGOs. In 1987, ACFOA commenced operation of an Appraisal and Evaluation Unit (in 2006, the Development Advice and Training Unit) to assist agencies in project design and appraisal, long-term planning and evaluation of projects (ACFID 2006).

The emphasis on evaluation of projects during the 1980s fostered concern amongst NGOs to ensure their projects were sustainable, and environmental issues were of increasing concern (ADAB 1983: 27). Mounting evidence that aid projects could be environmentally destructive and cause a new set of long-term problems for developing nations, along with increasing public concern in the developed world about environmental degradation, brought the environment firmly on to the agenda of most NGOs by the end of the decade. The strength of public concern about the issues, engendered in part by the activities of the conservation movement and environmental lobby within Australia, provided the focus for a major publicity campaign organised by ACFOA. The appointment of a new Minister for Foreign Affairs after a period of declining government commitment to aid provided the timing for the One World campaign launched in 1989. It focused on the interrelatedness of worldwide environmental destruction with other issues the voluntary aid movement had long sought to bring to public attention, including achieving sustainable development in harmony with the environment; more aid, fair trade and measures to lift the debt burden from developing countries; redirecting global arms expenditure to food, shelter, health and education for the world's poor; and the right of all people to participate in decisions that affected their lives (ACFOA 1989: 6). Working through the media and local community organisations, the campaign was marked by a high degree of unity and cooperation amongst Australian NGOs.

Thus, by the early 1990s, significant 'scaling up' of NGO advocacy had occurred, resulting in part from closer engagement with Australia's bilateral agency. Lobbying about the quantity of Australian ODA had been extended to include concerns with quality and impact. Major disaster appeals had given NGOs new media prominence and enabled them to develop cooperative working relationships with the media. As well, in environmental issues the voluntary aid community had found a new platform with which to reach the Australian public – one with popular appeal and growing respectability in the eyes of the public and the media. By the close of the twentieth century, advocacy was increasingly prioritised by Australian NGOs. The remainder of this chapter explores the extent and nature of the advocacy work of Australian NGOs in the last two decades.

Australian NGOs as advocates

Defined as 'lobbying, campaigning and any other means . . . to influence structures and policies that adversely impact on people living in poverty (but excluding development education programs in schools)' (Ollif 2003: 93), advocacy was at least a part of the work of 60 per cent of Australian NGOs by 2001 (from a sample of 65 NGO respondents to a questionnaire distributed to ACFOA members in that year (Ollif 2003)). Since the late 1970s, calls from the South had urged Northern NGOs to move into advocacy work; pressure to do so grew throughout the 1980s and strengthened throughout the 1990s (Durning 1989: 51; Verhelst 1990; Burnell 1991; Porter and Kilby 1996; Chapman and Fisher 2000; Ollif 2003: 31, 36–7). The earliest advocacy began among surveyed NGOs in 1945, and there was a clear trend in NGOs taking on advocacy during the last two decades of the twentieth century (see Figure 2.3). Thus, 46 per cent of the NGOs commenced advocacy in the 1990s, while only 28 per cent had done so in the four decades from 1940 to the end of the 1970s. Until the end of the 1970s, equal numbers of new NGOs engaged in advocacy immediately after their formation as did not. However, during the 1980s and 1990s, twice as many NGOs commenced advocacy immediately on establishment as did not, confirming its growing importance for NGOs during those two decades. Australian NGOs thus responded to calls from the South, their own members' requests and their own institutional learning about the importance of advocacy, by increasingly engaging in advocacy.

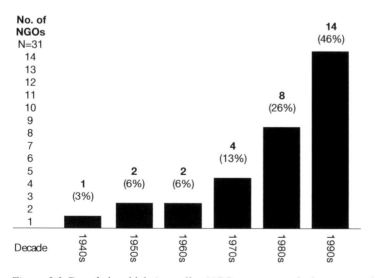

Figure 2.3 Decade in which Australian NGOs commenced advocacy work

The extent of participation in advocacy activities was closely related to staffing levels. Of NGOs not involved in any advocacy, most employed few, generally five or less, staff. An obvious exception was one Australian NGO with no global or international affiliation which, with 450 staff, did not participate in advocacy activities. The percentage of work time devoted to advocacy by NGOs was also revealing; approximately 12 per cent of NGOs indicated that they spent 76–100 per cent of work time on advocacy. However, some advocacy-only NGOs nominated a lower percentage, perhaps separating daily functions, like administration, fundraising and other activities, from advocacy work. All NGOs in the over 76 per cent category were formed between 1988 and 1995. While this growth in Australian advocacy-only NGOs could not be said to represent a significant trend, it does add weight to the UNDP's finding that 42.5 per cent growth had occurred internationally during the decade 1990–2000 amongst 'Law, policy and advocacy International NGOs'; numbers of such NGOs rose globally from 2,712 in 1990 to 3,864 in 2000 (UNDP 2002: 103). Even so, at the beginning of the twenty-first century, for the majority of NGOs in Australia, advocacy was only one aspect of their broader mandate.

This is borne out by analysis of the percentage of total NGO budget devoted to advocacy (see Figure 2.4). Most NGOs (90 per cent) indicated that time spent on advocacy and the funds allocated to it were approximately equal. Debate persists within the NGO community about the cost-effectiveness of advocacy and the proportion of resources that should be allocated to it (Roche and Bush

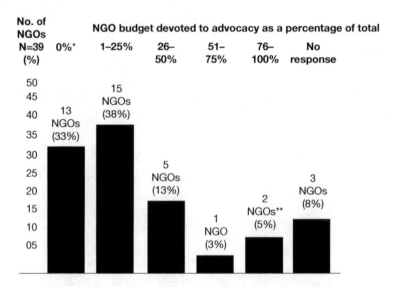

* The NGO does not allocate funds specifically to advocacy work

** As for Question 11 above, one of the 'no response' answers has been put into the 76–100% bracket.

Figure 2.4 Percentage of Australian NGO budgets devoted to advocacy

1997: 11); however, 65 per cent of Australian NGOs allocate resources directly to advocacy activities.

Prompted by criticism that a real weakness of NGO advocacy is an overall absence of clear strategy (Edwards 1993: 168), NGOs were asked whether they possessed a written advocacy policy. While 54 per cent of Australian NGOs active in advocacy had a written advocacy strategy or policy, nearly half did not, giving credence to Edwards's criticism. Of those with a formal advocacy strategy, the intent of many was embodied in the following objective of one NGO: 'To seek human development outcomes brought about through policy change by governments, action by social movements and deeper public understanding of social/political issues, at home and abroad.' Two of the 65 NGOs had campaign-specific objectives, but others included:

- to alleviate suffering worldwide;
- to be actively involved in ACFOA to influence government aid policy and maximise NGO effectiveness;
- to advocate at a state, national and international level and to raise public awareness;
- to ensure Australia lives up to its responsibilities;
- to create and nurture a network amongst supporters engaged in advocacy;
- to engage on one key issue every two to three years, integrating education, promotion, policy and advocacy outcomes for that issue;
- to be politically non-partisan and non-ideological;
- to achieve results while operating within an area of expertise; and
- to try cooperative approaches first.

Although not specifically writing about NGO work, Roche argued that 'poor institutional learning and weak accountability mechanisms are characteristic of many NGOs, which both lead to and are the result of the absence of professional norms and standards' (1999: 2). However, 77 per cent of Australian NGOs indicated that all staff understood well their advocacy objectives, suggesting considerable commitment to developing professional norms and standards in their advocacy work. A variety of methods was used to ensure staff familiarity with advocacy objectives, including in-house training, communication with field staff volunteers and boards of directors, campaign-specific or thematic meetings, evening training seminars and courses, staff participation in formulation and development of policies, position papers and publications, and clear advocacy strategies.

Australian advocacy campaigns

Some 87 per cent of respondent NGOs had participated in an advocacy campaign in 2001. Unsurprisingly, 'aid' was the most common focus of such campaigns, with 60 per cent of NGOs involved in a campaign related to 'aid' issues at some stage, including levels of aid, methods of delivery and 'quality' of aid. Thus, lobbying about official development assistance has remained an abiding concern of Australian NGOs. After aid was 'asylum/refugees', a very critical issue in Australia at the time (Mares 2003), with 52 per cent of NGOs engaged in a campaign relating to this issue, closely followed by 'debt relief', about which 50 per cent of NGOs advocated. Landmines, health and education, HIV/AIDS and protecting the environment were significant, as was country-specific advocacy, with 44 per cent of NGOs conducting advocacy related to 28 different nations. Other advocacy campaigns included campaigns about weapons and small arms; child soldiers, children's rights, street children and child labour; China's human rights record; consumer competition and privacy; corruption; development of an NGO disability network; energy; financial services; food and nutrition; globalisation; human trafficking; infant formula; the East Timor international criminal tribunal; international humanitarian law; IT/communications; microfinance; natural resources; poverty; sustainable agricultural practices; sustainable development; tobacco control; violence; the 'Vision 2020 Right to Sight' campaign; and water. In total, 42 different issues emerged as themes about which Australian NGOs have advocated at some time during the past ten years; despite obvious overlap between some, it is clear that a great diversity of development issues is of concern to NGOs as advocates.

Assessing the impact of advocacy initiatives is challenging for NGOs, for a variety of reasons, not the least of which is the difficulty of disentangling the impacts of advocacy from other influences. The most consistent criticisms of NGO advocacy though relate to the lack of deliberate and thorough evaluation of advocacy projects (Edwards 1993: 168; Roche 1999: 32; Anderson 2000: 450; Madon 2000: 4; Nelson 2000a: 486) and of NGOs learning from and responding to the evaluation of this work (Chapman and Fisher 2000: 151). These criticisms have some validity, for deliberate and thorough evaluation of advocacy was very limited, with only 18 per cent of respondent NGOs 'always' evaluating advocacy campaigns. This 'often' happened for a further 18 per cent, and for 32 per cent it 'sometimes' happened. Almost a quarter of respondent NGOs (24 per cent) 'rarely' evaluated their advocacy campaigns. NGOs may find the challenge of meaningful evaluation too great or, based on the common perception that donors are reluctant to finance advocacy (Sogge and Zadek 1996: 88), may try to reduce advocacy costs by limiting time and resources given to evaluation. Campaigns may be so different in focus that an evaluation of one campaign may not be usefully related to the next; however, surprisingly, only 34 per cent 'always' or 'often' used evaluation to inform future campaigns.

Alliances for advocacy

Over the last two decades, development NGOs have forged alliances with grass-roots organisations in the South, and also with other Northern development NGOs, including those with related missions – such as peace, the environment, women, human rights and consumer affairs. NGOs are generally better positioned to lobby governments for political change when working in alliances and networks, for these multiply the impact of individual NGOs (Welch 2001: 268) and thus, through them, NGOs can become 'a force for dramatic social change' (Edwards and Hulme 1992: 26). However, 'a failure to build strong alliances' was also identified as a weakness of NGO advocacy (Edwards 1993: 168).

Australian NGOs affiliated with a global alliance of NGOs were more likely to engage in advocacy than those with no such affiliation. Thus, 75 per cent of NGOs engaged in advocacy work were affiliated with a global alliance of NGOs or an international NGO; of NGOs which had never been involved in advocacy work, only 42 per cent had an alliance. The strong correlation between NGO alliances and advocacy activity suggests either that global alliances encourage members to join advocacy campaigns or that NGOs seek affiliation, seeing participation in a global alliance as beneficial to their advocacy work.

Ninety per cent of NGOs had participated in campaigns in conjunction with other Australian NGOs, demonstrating their belief that working in national, as well as international, alliances multiplies impact and better places them to lobby for political change. Of these, 44 per cent mentioned working with ACFOA (to be expected given that they were all members of ACFOA). In particular, networks, most specifically the Jubilee 2000 campaign network, were mentioned by 37 per cent of NGOs. Smaller NGOs preferred to work with large NGOs, even though the larger NGO made no mention of the smaller NGO. Larger NGOs mentioned work with ACFOA members, which would cover smaller NGOs with which they have had alliances; otherwise, large NGOs mostly listed ACFOA, Jubilee, and other large NGOs as advocacy partners. Evident therefore was the role that the larger NGOs (for example Oxfam, CARE and World Vision) played in creating alliances with the smaller NGOs. In addition to advocacy campaigns run by ACFOA, alliances with larger NGOs undoubtedly helped smaller organisations with limited resources to invest in meaningful advocacy.

Themes of advocacy campaigns conducted within alliances included debt (38 per cent of NGOs), followed by 'aid' (27 per cent). Opposition to landmines was a strong theme; other campaigns concerned refugees, violence against women, the environment, education, HIV/AIDS, blindness prevention and children's rights. 'Debt' was the most popular issue that NGOs advocated on in conjunction with other Australian NGOs, which is not surprising, for the global Jubilee 2000 campaign, which many Australian NGOs supported, had unprecedented success in mobilising thousands of NGOs worldwide. Since 1990, most agencies had worked with several alliances, with Jubilee 2000 again mentioned most often. Unsurprisingly, Australian NGOs collaborated with the international body to which they belonged, so, for example, CARE Australia specified CARE International. Many also collaborated with large international NGOs such as Greenpeace, Oxfam or World Vision. Others worked with NGOs with a common rationale; for example, those concerned with health-related issues worked together on campaigns reflecting their focus. However, one-third fewer NGOs collaborated in a global advocacy alliance than in an alliance with an Australian NGO.

Fewer Australian NGOs (54 per cent) engaged in an alliance with a Southern NGO than in global alliances or with other Australian NGOs. Information received from Southern partners legitimates advocacy, yet many NGOs did not take advantage of alliances with Southern NGOs. While some NGOs used Southern project partners to provide information, they did not work with those project partners in alliances for advocacy purposes. Themes of advocacy conducted in conjunction with Southern partners paralleled those which were the focus of advocacy involving other forms of alliance, with the most commonly mentioned again being debt, HIV/AIDS and landmines.

Official funding and advocacy

Considerable debate within the development community concerns the possible adverse impact of donor-imposed conditions on the independence and legitimacy of NGOs (see Edwards and Hulme 1995; Smillie 1995, 1999b: 9–11; Raffer and Singer 1996: 142; Weiss and Collins 1996: 47; Chambers 1997; Hailey 2000: 404; Malavisi 2001: 54). This debate is lively amongst Australian NGOs. AusAID has been the 'largest single contributor to the Australian development NGO community' (AusAID 1999: 3). Sixty-two per cent of NGOs said they received funding from AusAID; 8 per cent not receiving AusAID funding in 2001 had done so in the past. Thirty per cent of NGOs had never received AusAID funding. Interestingly, half of the NGOs which had received AusAID funds had also advocated against an AusAID policy; exactly the same proportion of NGOs not having received AusAID funds had also advocated against an AusAID policy at some time. Therefore, receipt of AusAID funds does not necessarily impact on an NGO's propensity to advocate against AusAID policies, despite concerns that receipt of government funding could compromise NGOs commitment to lobbying.

The majority of NGOs did not think that AusAID denied funding to those which advocated against government policy. Only a minority believed, or were prepared to admit, that AusAID denied funding to those NGOs which advocated against it, although NGO staff may have been reluctant to report this on a questionnaire where the name of the NGO was generally disclosed, especially if the NGO was then receiving AusAID funding. On the other hand, at least for one respondent who had considerable experience of AusAID, agencies funded by AusAID *do* feel constrained in their ability to criticise AusAID. However, such constraints are subtle; NGOs may well self-censor their advocacy in the face of potentially powerful government opposition, where their success is likely to be limited.

NGOs expressed concern that AusAID does not fully appreciate the contribution NGOs have made to development and to capacity building within national governments. On the other hand, one NGO staff member commented that AusAID well understands the role of NGO advocacy but 'probably doesn't like it as it would often expose their shortcomings in foreign policy and aid'. Another reported that 'the Director-General of AusAID has openly told the

ACFOA Council that they should not criticise AusAID'. Yet another was more understanding of AusAID's position, admitting that, as AusAID is part of the Department of Foreign Affairs and Trade, even if it is empathetic towards NGO advocacy it may be constrained by diverse government policies. Thus, even where AusAID itself recognises NGO demands, it is often unable to support them. The relationship between Australian NGOs and government thus remains complex, but NGO concerns about the quantity and quality of Australian ODA remain central to their advocacy activities.

Across the generations

General literature about NGOs has documented that, in every decade since the 1940s, and particularly in the last two decades, advocacy has increased worldwide as an NGO activity. This chapter has demonstrated that, in the Australian context, this has certainly been the case. While the genesis of the Australian NGO movement lay in the formation of missionary organisations and charities, these agencies were transformed into deliverers of project aid; new organisations were formed and became increasingly politicised. Increased emphasis on education and lobbying during the 1980s resulted in a dualism in agency functions. Throughout the same decade, growing cooperative working relationships with the media, and increased cooperation and coordination of NGO activities culminated in NGOs working together in joint advocacy campaigns in the 1990s. Forty-five per cent of Australian NGOs commenced advocacy work during the 1990s, it has become increasingly prioritised by many development NGOs in Australia, and advocacy-only NGOs are growing in numbers in Australia. Thus, NGOs have metamorphosed; no longer acting simply as donors of practical assistance to communities in developing nations, they are now active advocates seeking to change national, regional and global policies and structures which perpetuate poverty.

The exploration of advocacy here demonstrates that a number of key issues face Australian NGOs. These are representative of those facing Northern NGOs in general: many do not conduct careful and systematic evaluations of their work; alliances have become increasingly important to NGO advocacy strategies in the last two decades; the relationship between NGOs and official donors which increasingly fund their work and the impacts of this on their advocacy activity are debated; and the failure of NGOs to engage in advocacy activities with Southern partners is surprising given that such relationships should lend legitimacy to NGO advocacy activities. The following chapter explores further the growth in, and nature of, the advocacy work of NGOs, with a focus on the advocacy work of a number of Australian NGOs.

3   Speaking out

Australian NGOs as advocates

Cathryn Ollif

From the early 1990s, advocacy increased in importance for Australian NGOs. As outlined in the preceding chapter, during this time many existing NGOs began advocacy and many new NGOs were established with advocacy either as part of their mandate or as their sole *raison d'être*. This chapter explores these developments, presenting some aspects of research, conducted between 2000 and 2003, into the advocacy work of Australian NGOs. In particular, it presents case studies of six NGOs and is based on interviews with past and present staff from the selected NGOs, personnel from the Australian Council for International Development (ACFID) and representatives of AusAID, and an exploration of existing published documents. For five of the six case study NGOs, advocacy was either initiated or significantly scaled up in the 1990s.

Introducing the NGOs

Varied organisational characteristics were used as a basis for selecting NGOs for case study, including that: some were secular and others were faith-based NGOs; in terms of numbers of employees they varied in size; four were established before 1990 and two founded more recently; three engaged in advocacy before 1990 and three committed to advocacy activities after that date; and some were 'special interest' NGOs while others were general 'development-focused' organisations. As the following brief histories of the NGOs reveal, diversity also existed in the extent to which each NGO allocated funds specifically to advocacy, whether advocacy was their sole focus and whether and to what extent they have engaged in alliances for advocacy purposes.

Anglicans Cooperating in Overseas Relief and Development

The first and smallest among the case studies, Anglicans Cooperating in Overseas Relief and Development (*Angli*CORD), is a small,[1] faith-based NGO, located in Melbourne, which was established in 1987. *Angli*CORD initiated advocacy work in 1994 and has participated in campaigns on aid, asylum seekers/refugees, debt relief, HIV/AIDS and landmines. At the end of the 1990s, advocacy became much more important to the organisation. The *Angli*CORD Board decided to significantly increase its advocacy activities and employed a chief executive with the experience and contacts to assist the organisation to do so. The chief executive officer believed that, given its size, *Angli*CORD's most useful advocacy work occurred as part of alliances working with other NGOs, to which it brings, among other experience, particular expertise on Africa and HIV/AIDS.

The Australian Reproductive Health Alliance

The Australian Reproductive Health Alliance (ARHA) is a small, secular NGO, which has engaged in advocacy since its establishment in 1995. It was one of several NGOs created globally following the United Nations Population Fund's International Conference on Population and Development (ICPD), held in Cairo in 1994. A major part of ARHA's work is monitoring the Australian government's response to the ICPD Program of Action.[2] ARHA sees its mandate as promoting public support, both within Australia and internationally, for improvement in the wellbeing and status of women and the development of reproductive health in families and individuals. ARHA considers itself to be essentially an 'advocacy-only' NGO. Its advocacy has included being part of campaigns on aid, asylum seekers/refugees, debt relief, health and education, HIV/AIDS, reform of international institutions, and the Tobin tax.

The Foundation for Development Cooperation

The Foundation for Development Cooperation (FDC) is not a development NGO in the usual sense of engaging in relief and project work. Rather, FDC's work focuses exclusively on an integrated mix of strategic research, policy development and advocacy, its goal being to achieve sustainable development and a reduction in poverty in Asia and the Pacific. Thus, while not a 'traditional' aid agency, FDC shares the same goal as all aid and development NGOs, that is, a reduction in poverty. Since its establishment in 1990, this small organisation has engaged in advocacy relating to aid, promoting corporate responsibility, microfinance, protecting the environment, reform of international trade rules, and the Tobin tax.

The International Women's Development Agency

In 1985, at a time of growing awareness that women were often neglected in development processes, the International Women's Development Agency (IWDA) was established. A mid-sized NGO, it has, since its establishment, engaged in some advocacy work. Its advocacy campaigns have included Aboriginal Australia, aid, asylum seekers/refugees, corporate responsibility, debt relief, protecting the environment, reform of international institutions, violence to women, trafficking in women, and poverty reduction.

Historically, the type and amount of advocacy undertaken by IWDA depended on the priorities held at the time by the chief executive of the NGO. In 1993, it significantly scaled up its advocacy when a chief executive with an activist background joined the agency for two years. After that individual left in 1995, IWDA's advocacy work declined. Following an extensive strategic review of the aims and objectives of the organisation and its supporters in 2001, IWDA concluded the NGO was too small to be successful at stand-alone advocacy and discontinued it. The (then) chief executive believed that IWDA could most usefully engage in advocacy as part of ACFID alliances, specifically where it could lend its expertise to influence gender and development discourse and practice in Australia and internationally. In late 2003, IWDA appointed a new executive director who increased cross-stakeholder dialogue as a form of advocacy for gender issues, to stimulate discussion in the aid and development sector about gender issues amongst other NGOs, government departments, the commercial sector, bilaterals and multilaterals, academics and practitioners.

Oxfam Australia

Initially established in 1954 as the Food for Peace Campaign, and later known as Community Aid Abroad (CAA),[3] what is now known as Oxfam Australia is today, through its membership in Oxfam International, part of a 'super-NGO'.[4] From the outset, CAA engaged in some advocacy and since the 1970s the NGO has been a leader in NGO advocacy in Australia. Today, most of Oxfam Australia's national and international operations are coordinated from its head office in Melbourne, and offices in most states supervise day-to-day activities, both state and local. Its advocacy campaigns have included Aboriginal Australia, aid, asylum seekers/refugees, core labour standards, corporate responsibility, debt relief, East Timor, health and education, HIV/AIDS, reform of international institutions, reform of international trade rules, and globalisation.

Oxfam Australia has engaged in advocacy for much longer than most NGOs in Australia. In the mid-1970s, supporters split ranks over the NGO's activism regarding Indonesia and East Timor. As a result of that crisis, members who did not wish the NGO to be political left the organisation but many new members joined precisely because they were looking for an activist organisation (Blackburn 1993; Rugendyke 1994: 185). Following restructure in 1990, CAA committed itself to more intensive coordination and development of its campaigning activities. In 1996, CAA was one of nine independent NGOs which became founding members of Oxfam International. Members of Oxfam International share common values, as well as educational and campaigning resources, and cooperate in development programmes and disaster relief, and membership has enabled Oxfam Australia to greatly increase its advocacy capacity. An Oxfam Australia advocacy manager commented that the role of advocacy work is 'every bit as important' as any other work of the NGO.

World Vision Australia

World Vision Australia (WVA) was established in 1967, but did not involve itself in advocacy for its first 20 years. As a member of World Vision International, WVA was an internationally affiliated NGO long before it commenced advocacy 'informally' in the late 1980s and 'formally' in 1990. In the late 1980s, at the request of some field staff, WVA began a little unstructured advocacy work. In 1990, following increasing requests from staff, WVA adopted the newly formulated World Vision International policy that all World Visions should begin a formal programme of advocacy. Its budget is the largest of the development NGOs in Australia and, as a member of World Vision International, it is in the 'super-NGO' category. Its advocacy campaigns have included Aboriginal Australia, aid, child labour/child soldiers, debt relief, HIV/AIDS, landmines, protecting the environment, reform of international trade rules, small arms, corruption, and human trafficking. WVA continues to see its Child Sponsorship programme and overseas project work as the core work of the NGO, with advocacy playing an important 'supporting' role.

Advocacy in practice

The role of advocacy was described as the 'strategic use of information to . . . improve the condition of those living in poverty' (Roche 1999: 192). NGOs use a range of activities including campaigns, lobbying governments and other stakeholders, education and research to try to create an aware public in their own countries, and to influence structures and policies nationally and internationally to bring about change in the interests of eradicating poverty and its underlying causes. Each of the NGOs used all, or a combination of, these activities. Advocacy work may be highly visible, requiring extensive in-house resources, as is the case for Oxfam Australia and WVA, or, when resources are limited, the NGO may complete most of its advocacy within networks, as is the case for *Angli*CORD and IWDA.

Each of these NGOs interpreted the role of advocacy in its own way. For example, FDC does not involve itself in either public campaigning or direct lobbying of governments or other stakeholders but uses advocacy to communicate its research and policy work to decision makers in government and business, academics, other NGOs and community leaders, within Australia and overseas. Likewise, for ARHA, advocacy is a way of communicating its research to inform parliamentarians, government departments, students and others about issues related to reproductive health. ARHA sees the role of advocacy as a circular process which never really ends, '. . . ensuring equity for those in no position to fight for it themselves, whilst giving them the tools to increasingly take over the task' (Proctor 2000a: 6). ARHA has engaged in some innovative advocacy, including its All Party Parliamentary Group study tours. These tours were devised to raise awareness of participating parliamentarians about the issues surrounding reproductive health. Tours, which have visited Fiji, Vanuatu, Thailand, Vietnam and the Philippines, have been among ARHA's most successful advocacy work, and continue subject to availability of funds. ARHA is also active in campaigns and will lobby parliamentarians on policy matters if necessary.

The role of advocacy for Oxfam Australia has been explained in promotional literature as a process of '. . . changing hearts and minds – and government and company policies – [and] is as critical to our aims as our most grassroots development programs' (CAA 2000a: 6). In contrast, although WVA has developed an active advocacy department, it continues to see advocacy as playing an important *supporting* role for its Child Sponsorship programme and overseas project work.

If an NGO decides to scale up its advocacy work, it risks alienating those supporters who are not in favour of the shift in emphasis in the role of advocacy within the NGO. This is not a consideration for NGOs like FDC and ARHA, which were established with a clear advocacy focus from the beginning. However, IWDA, Oxfam Australia, WVA and *Angli*CORD must consider supporters' views when contemplating any serious shift in direction, or risk losing supporters. After Oxfam Australia both lost and gained a significant number of supporters because of its political activism in the mid-1970s, it occasionally surveyed supporters to determine whether they were satisfied with the NGO's direction. In 2000, for example, a survey was used to confirm that Oxfam Australia's supporters wanted the NGO to continue to advocate strongly for social justice.

When WVA first gave advocacy a greater role, it found many of its supporters were initially sceptical about the NGO involving itself in political work, reflecting the conservative Christian ethos of the organisation and its constituency. Many WVA supporters preferred to see the bulk of their donations go to overseas programmes so, to gain support for advocacy work, WVA staff embarked on a programme to educate supporters to accept that advocacy work had become necessary to successfully fight poverty. The NGO stressed that its extensive experience in the field gave the agency valuable knowledge about fighting world poverty that should be shared with policy makers. WVA believes its constituency now largely agrees with its advocacy role, although support tends to be provided in the form of responses to requests for petition signatures and letter writing, rather than financial support specifically for advocacy work. However, whereas about 70 per cent of donations support the Child Sponsorship programme, less than 5 per cent of donations go to WVA's Campaign Partners programme of monthly donations to support advocacy.

In *Angli*CORD's case, donors also had to be convinced to support the Board's decision to scale up the role of advocacy within the NGO. *Angli*CORD's constituency is almost entirely made up of Anglican congregations across Melbourne and Victoria and, as is the case for WVA, donors prefer to see as much money as possible used for projects overseas. However, supporters were receptive to *Angli*CORD's arguments in favour of advocacy.

Four of the case study NGOs run programmes directly aimed at educating youth. Both Oxfam Australia and WVA believe it is important to educate youth to be concerned global citizens and both have campaign programmes aimed at younger supporters and provide resource materials for school teachers. *Angli*CORD has a programme dedicated to producing information in a youth-friendly format, which is widely distributed to high schools. ARHA's innovative work with youth includes visiting schools in capital cities to run 'youth seminars'.

The experience of these NGOs confirmed that the leadership of an NGO can play an important role in the NGO's commitment to advocacy (Hodson 1992: 127). For IWDA it was the chief executive who had driven its advocacy. In the case of the 'advocacy-only' NGOs, the chief executives of both FDC and ARHA were employed by the NGOs in part at least because of their extensive backgrounds in policy and advocacy work. When the *Angli*CORD Board decided to significantly increase its advocacy work the Board sought and employed a chief executive with the experience and contacts to do this. When Oxfam Australia needed a new executive director in 2001, the advocacy manager moved into the position, confirming Oxfam Australia's commitment to advocacy.

Most NGOs believed that cooperation within alliances, not competition between NGOs, was the key to achieving common advocacy goals. NGO staff welcomed the possibility of advocacy raising the NGO's profile but stressed that such work was not used to 'compete' with other NGOs for funds. Indeed, because many donors supported more than one NGO, organisations cooperated rather than competed with each other. NGOs believed that, to lessen any perception that advocacy work was little more than self-promotion, NGO campaigns should be supported by rigorous research and high ethical standards.

Confrontation or cooperation?

While some NGOs are confrontational in their advocacy, others prefer to conduct an 'insider' debate, and NGOs may adopt both approaches at once. Advocacy that attempts to influence processes, structures or ideologies – that is, seeking *fundamental change* – is more likely to be confrontational. Advocacy attempting to influence specific policies, programmes or projects – that is, seeking *incremental reform* – is likely to be based on cooperation rather than confrontation (Edwards 1993: 164). Unsurprisingly, it is the INGOs and ACFID networks that most often tackle fundamental change because this can take a long time and use extensive resources. When NGOs use advocacy to seek incremental reform by influencing specific policies or projects, it is likely to be in areas that are important to their own interests.

Most of the NGOs have used both approaches with respect to their advocacy, except FDC, which never takes a confrontational approach, preferring to always work cooperatively with the targets of its advocacy. WVA also prefers cooperative, non-confrontational advocacy, but is prepared to be confrontational if necessary and has been involved in confrontational campaigns. ARHA's advocacy is a mix of both styles – it frequently challenges the government to honour its commitments to various United Nations agreements but its overseas tours for parliamentarians have, according to several past participants, generated an enormous amount of goodwill and heightened understanding by parliamentarians about ARHA's work. Oxfam Australia believes it has a reputation for being frequently confrontational with government and is willing to engage in confrontational advocacy if necessary, but it also works hard at building relationships through cooperative advocacy.

Both Oxfam Australia and WVA's advocacy departments actively try to cultivate contacts within relevant government departments in order to maintain an ongoing 'insider' dialogue with decision makers. The patient building up of personal contacts can be an influential type of advocacy (Holland and Blackburn 1998: 3) and, for all case study NGOs, where possible, building personal contacts with targeted policy makers is important.

Oxfam Australia increased its ability to advocate for fundamental change when it became part of a global partnership. WVA, on the other hand, had been a member of a global partnership for some time when it took up advocacy work and was thus well placed to advocate for fundamental change. The Oxfam International and World Vision International partnerships work hard to build reputations for producing credible and well-researched evidence as a basis for their advocacy work. Both have, at times, gained access to the highest level of development decision making in the United Nations and other agencies in the multilateral system, as well as access to national policy-making institutions. When the smaller NGOs are involved in seeking fundamental change, it is usually as part of an advocacy coalition. Several of the NGOs believed that ACFID networks were the ideal way for a small NGO, with limited advocacy resources, to be involved in campaigns which attempt to influence global-level processes, structures and ideologies.

Tools of NGO advocacy

Regardless of whether NGOs are seeking incremental reform or fundamental change, for advocacy to succeed it is vital that NGOs can demonstrate public support when they lobby politicians (Burnell 1991: 250). All the case study NGOs, with the exception of FDC, actively solicited demonstrated support during campaigns, often in the form of signed postcards, petitions, e-mails or letters directed to politicians.

Petitions are an increasingly popular way for NGOs to demonstrate that they have public support. WVA collected a quarter of a million signatures for a petition, which it believed played a key role in the success in Australia of the landmines campaign. Since the mid-1990s, NGOs have lobbied for banning the use of landmines in war, recognising that landmines buried in 80 nations are responsible for an estimated 20,000 casualties each year (ABC 2006). The largest petition so far presented to an Australian government on a foreign policy issue was the Jubilee 2000 petition. The purpose of the petition was to draw attention to the debt crisis and to develop mechanisms to offer debt relief to heavily indebted nations unable to meet their debt repayments. It ultimately gathered 443,803 signatures nationally; these signatures were added to over 24 million signatures collected worldwide and presented at the United Nations Millennium Summit in New York in September 2000 (Jubilee 2000 2000: 5).

Advances in communications technology have given advocacy an immediacy in national and international reach that did not exist before the 1990s. Using the internet, information and experiences are now easily shared, and events around the globe can be transmitted within minutes, keeping activists up to date at minimal cost. Coordination of lobbying efforts is now possible at short notice, and the use of private 'intranet' communications allows for the internal management of global campaigns. The Jubilee 2000 campaign for debt relief became the benchmark for what could be achieved using communications technology during a global campaign and taught NGOs much about effective campaigning and the power of appropriate media and internet use. Dozens of NGOs, church groups and other organisations across Australia joined thousands of similar organisations around the world for what *Angli*CORD described as 'the most successful campaign that we have been a part of'.

NGO websites keep campaign supporters up to date about campaigns. All the NGOs have websites giving information about their current campaigns. Both Oxfam Australia and WVA's websites provide extensive information about their active campaigns; supporters can lend instant support to the campaign of their choice by signing online petitions or sending online letters to political leaders in support of specific issues. The global response to such petitions confirms the potential power of global campaigns. For example, Oxfam International's Make Trade Fair campaign received over 6.5 million signatures worldwide on its Big Noise petition and, in 2005, Oxfam Australia's website invited people to sign the Make Poverty History campaign petition and thus join '. . . the 8 million people around the world who have already signed the petition' (Oxfam Australia 2005).

Websites have thus become an essential tool for NGOs' advocacy work. In late 2000, Oxfam Australia claimed that its website was '. . . the most popular humanitarian site in Australia, and the fourth most popular globally amongst Australian net users. . . . In the general category we rank eighth amongst Australian sites' (CAA 2000b: 3). At that time, Oxfam Australia had some 1,400 pages of information about its programmes, events and campaigns on the website, with more being added on a regular basis. Oxfam Australia could no longer envisage involvement in advocacy without the use of the internet to harness the help of its supporters.

Alliances for advocacy

A common criticism of advocacy is that NGOs fail to build strong alliances for advocacy purposes (Edwards 1993: 168). Data from the research questionnaire revealed that, in Australia, 90 per cent of NGOs had worked in an alliance with other Australian NGOs for advocacy purposes and 60 per cent had also worked in a global alliance, though these figures say nothing about the utility of the alliances or whether the NGO will attempt to further build alliances or make full use of alliances in the future. However, NGO staff were aware of the value of creating strong and dynamic alliances for the purposes of advocacy and most actively sought stronger and more diverse alliances, particularly within Australia, for future advocacy. In unity there was strength.

From the late 1980s, both Northern and Southern NGOs increasingly joined networks and coalitions specifically for advocacy purposes (Korten 1990: 93). The rapid advances in communications technology of the 1990s were instrumental in enabling many new national and international NGO networks to develop. Alliances for advocacy purposes multiplied the impact of individual NGOs. Through strength in numbers, alliances were better positioned to lobby governments for political change than NGOs working alone. Members of a network can share expertise, knowledge and resources. Each of the NGOs, except FDC, which preferred to maintain independence to allow it to speak freely on issues, has engaged in advocacy as part of alliances in Australia. All six NGOs worked in advocacy alliances with Southern NGOs, usually with project partners or NGOs with similar interests. All believed their alliances with Southern NGOs are important for the success of their advocacy because, by providing 'on-the-spot' information and knowledge, they provide legitimacy to the advocacy.

The importance to NGO advocacy of ACFID networks was stressed by the NGOs. It is unsurprising that, as members of ACFID, most have been part of ACFID advocacy networks but, for all except FDC, ACFID played a very important part in their advocacy. Four of the case study NGOs had staff on the ACFID Advocacy and Public Policy Committee, including *Angli*CORD's chief executive officer, who chaired the Committee, ARHA's chief executive officer, WVA's advocacy manager and an Oxfam Australia staff member from its head office advocacy department. More recently, Oxfam Australia's executive director has become ACFID's vice-president.

While FDC did not participate in ACFID campaigns, it recognised the importance of those alliances for most NGOs and, through sharing its research and policy work, FDC's work often informed ACFID's advocacy. FDC seminars were usually attended by at least some ACFID advocacy staff. Thus, FDC contributes to ACFID campaigns, perhaps quite significantly, although largely invisibly.

ARHA, the other 'advocacy-only' case study NGO, saw alliances in Australia and overseas as an extremely important part of its advocacy work. Similarly, alliances were an important advocacy tool for *Angli*CORD which, as a small NGO, believes it benefited from the shared resources of alliances, while bringing to the alliance its own expertise. For IWDA, alliances were the most viable way for it to engage in advocacy in Australia and it will continue to lend its expertise to networks where it believes it can contribute positively.

Both Oxfam Australia and WVA work closely with ACFID on all their advocacy work. However, it is their global partnerships – Oxfam International and World Vision International – which provide the Australian NGOs with their most important alliances for advocacy work. The importance of these global partnerships for the advocacy work of these INGOs cannot be overemphasised.

When the Oxfam International partnership was created in 1995, Oxfam Australia (then Community Aid Abroad) and other members brought to that partnership their own long histories of advocacy. Oxfam UK, particularly, has a long-standing and well-respected tradition of advocacy work. As Community Aid Abroad, Oxfam Australia was already seen as the leader of advocacy work in Australia but, as a result of joining Oxfam International, its profile increased as an activist NGO. In contrast, the various World Visions brought virtually no previous advocacy experience to World Vision International when it commenced advocacy work in 1990, but the global alliance soon made its mark among advocates for social justice, through the power of substantial staff numbers and resources.

During the 1990s, one significant difference between Oxfam Australia's and WVA's use of alliances for advocacy purposes was that Oxfam Australia's advocacy was focused and it usually ran the campaigns it was involved in, with other NGOs joining. In contrast, as a newcomer to advocacy work, WVA had a strategy in its first decade of advocacy work of providing support to campaigns of other NGOs. Towards the end of the 1990s, as it became an experienced campaigner, WVA began to run some campaigns itself, gradually moving towards the Oxfam Australia model of deciding the focus of its advocacy and then leading campaigns. It believed increased clout, from diverse ethnic, gender, age and ideological groups, was one of several good reasons for seeking out broad-based coalitions for advocacy work (World Vision Australia 2002: 8).

It has been claimed that a downside of alliances is that communication and collaboration between international NGOs working in alliances has often been poor (Madon 2000: 5). However, staff from both Oxfam Australia and WVA rejected this notion in relation to work with other Australian NGOs, though both suggested that in working with INGOs their own NGO may have been guilty of poor communication at times. However, Australian INGOs may be their own harshest critics in judging their performance within alliances, because other NGOs saw only the benefits of working with INGOs in terms of resources, reputation, shared knowledge and the learning experience. Of course, NGOs belong to a small community in Australia so it may be that they were not prepared to be critical of the INGOs, seeking merely to work more effectively with them.

Advocating through the media

Over the last two decades, use of the media has become an important tool for successful NGO advocacy (see, for example, Burnell 1991: 17; Simon 2003: 9). Good media coverage not only educates and informs the public and politicians, but can also prove to NGO supporters that the NGO 'is taken seriously on its positions' (Lindenberg and Bryant 2001: 194). The Jubilee 2000 campaign broke new ground in its extensive use of the media to build a global

It's a matter of life and debt.

ACT NOW FOR JUSTICE by signing this postcard and sending it back to Jubilee Australia. We will forward it to the Federal Treasurer, Peter Costello urging him to cancel the debts owed to Australia by very poor nations and to use his position in international forums to advocate for a fair debt arbitration process between rich and poor nations.

Signature _____

Name _____

Address_____

In the world's most impoverished nations, the majority of people do not have access to clean water, adequate housing or basic health care. These countries are paying debt service to wealthy nations and institutions at the expense of providing these basic services to their citizens. If the debts of these countries were cancelled and used wisely, the lives of thousands of children around the world could be saved.

The Prime Minister has committed Australia to the Millennium Development Goals, which aims to halve world poverty by 2015. This will not happen without large scale debt cancellation. Therefore, Jubilee Australia is urging the Federal government to stop collecting debts from some of the world's poorest countries including Vietnam, Indonesia, Bangladesh, Nepal, The Philippines and Papua New Guinea. It is unjust that Australia continues to collect these debts.

The debt system also needs to be changed. Jubilee has called for an international insolvency framework that includes a just and transparent way of managing debt crises. Such a framework would allow poor nations to have an equal say in negotiations with lenders.

Find out more about the campaign and how you can help. Visit www.jubileeaustralia.org

JUBILEE AUSTRALIA
drop the debt

PLEASE AFFIX STAMP

The Treasurer

C/- Jubilee Australia

Locked Bag 199

Sydney

NSW 1230

Figure 3.1 Australian Jubilee 2000 postcard (front of postcard, opposite; back, above)

campaign. Effective use of print and visual media, as well as of the internet, enabled Jubilee 2000 to educate millions of people around the world about the debt issue and the impact of indebtedness and structural adjustment policies on the poor (Walker 2000: 18).

Indicative of the importance INGOs place on the relationship between the media and successful advocacy, both Oxfam Australia and WVA run well-resourced media departments within their organisations. WVA has a team of experienced journalists working in its media relations department in Melbourne to coordinate news media campaigns in conjunction with WVA's advocacy initiatives and provide support for other World Vision work. The media unit which Oxfam Australia created in 1989 has played an important role in Oxfam Australia's advocacy work ever since. The unit is committed to producing quality, timely news releases about Oxfam Australia's campaigns, and this has helped to establish Oxfam Australia as a legitimate source of information about social justice and human rights issues. Oxfam Australia considers good use of the media the most important aspect of successful lobbying of politicians, because politicians use the media to judge both what issues are important and whether the views of aid agencies enjoy general support in the electorate (OCAA 2002a: 22). WVA also used the media to find out what is on the government's agenda and then offers policy analysis based on its experience of that agenda item. WVA argued that a critical factor for successful advocacy was to try to influence government policies when the issue is *already* on the government's agenda. However, since the February 2004 appointment of the new chief executive, who is the brother of the Australian government's Treasurer, NGOs believe WVA is now well placed to do some 'agenda creating' of its own.

FDC actively sought to build a reputation with the media as a reliable and informed media source and employed a programme officer who devoted approximately 40 per cent of her time to media liaison. FDC believed the media were receptive to stories about the developing world if the information was from a reliable source and presented in the required format. ARHA also used the media extensively to give issues surrounding reproductive health as much exposure as possible, and frequently released its research as press releases, subsequently displaying these on its website. Oxfam Australia and WVA also made press releases readily available by placing links to their media work in a prominent position on the home page of their websites.

Legitimacy for advocacy

The increased profile of NGOs in the media has, at times, led to questions about the legitimacy of NGOs to speak on behalf of people living in other countries. Northern NGOs have frequently been criticised for not being representative of those on whose behalf they advocate – that is, the poor and marginalised of the South (Hudson 2000a: 91).

Oxfam Australia 'grappled' with the question of legitimacy during the 1990s, but a survey of its members in the late 1990s confirmed that members endorsed the agency's claim to speak on behalf of those who lacked the power and resources to speak for themselves. Members saw Oxfam Australia's legitimacy to campaign on behalf of people in the developing world as based upon its development work in the field. Other NGOs argue that their legitimacy is similarly based. Project and programme work in various countries of the South provides first-hand experience of the problems encountered by the people on whose behalf they advocate, so all NGOs attempt to establish a clear link between their advocacy work and practical experience. Influence, therefore, can be exercised with some degree of authority, legitimacy and credibility. Among the six NGOs, *Angli*CORD, IWDA, Oxfam Australia and WVA have all been pressured by Southern partners to advocate on their behalf. All six NGOs worked in alliances with Southern partners and stressed the importance of these in allowing them to access reliable information about key advocacy issues. Without such alliances, they would only be 'making assumptions'. Strong alliances with Southern NGOs, therefore, were critical in informing their research, policy development and advocacy.

Both Oxfam Australia and WVA believe the arguments they present in campaigns are now scrutinised more carefully by government and public institutions than ever before. Indeed, at the time of the research, the Australian press reported that a right-wing think-tank 'has had an ongoing research project devoted to the cataloguing and investigation of the NGOs', claiming that 'the increasing power of non-governmental organisations' is of concern to the current Liberal–National Party coalition government (McGuinness 2003: 3).

Criticism of NGO advocacy in Australia is not new, but NGO staff believe it is intensifying. In 1997 Oxfam Australia was criticised by the federal government over its prominent role in debates around Native Title and racism.[5] The government questioned Oxfam Australia's legitimacy to speak on behalf of indigenous Australians, but Oxfam Australia argued that non-indigenous Australians have a right to support indigenous people in their struggle for a fair deal on Native Title. Substantiating its claim, Oxfam Australia received its largest-ever response to a request for supporters to sign and return postcards for their campaign on Native Title. For Oxfam Australia, increased scrutiny indicates its arguments are being taken more seriously. World Vision also believes more public scrutiny is a direct consequence of its increased public profile through advocacy. Demands for accountability will continue to increase as the NGO advocacy profile increases. To meet these demands, NGOs must be able to demonstrate that their advocacy work is well researched, accurate and linked to first-hand experience, and has the support of its constituents and Southern partners.

Evaluating advocacy

Opinions differed between NGOs about the success of advocacy. ARHA was the most disillusioned; one senior staff member was unconvinced that it had been particularly successful and stressed that advocates must become braver and more outspoken at both national and international levels for advocacy to be successful. FDC also believed NGOs must become better at articulating their case to be successful.

Although the NGOs admitted their advocacy sometimes did not meet expectations, it remained worth the effort simply for awareness raising. Several recalled that their advocacy work had, at times, drawn criticisms from government, other organisations, the media, the public and even other NGOs, but they believed that the possibility of criticism must never stop them from speaking out, since criticism inevitably accompanied engagement in social justice advocacy.

Within the NGOs themselves, an agreed common criticism of their advocacy was that it could be of a higher standard with more resources and time, but then there is never as much time or money for the advocacy or other work as they would like. In Oxfam Australia's case, over the last few years, some staff members have been critical of the short time-frames committed to campaigns, which have generally run for between one and three years.

The most consistent criticism of advocacy work in the literature concerns the lack of deliberate and thorough assessment of advocacy projects (see, for example, Edwards 1993: 168; Roche 1999: 32; Madon 2000: 4). As NGOs have increasingly invested resources into advocacy over the last decade, greater pressure has been put on them by donors and other stakeholders to better evaluate the effectiveness of their advocacy work (Coates and David 2002: 534). Each NGO indicated it was trying to be more rigorous about the evaluation of advocacy projects; all agreed evaluations were important tools to assess what works, but also agreed that evaluations were both difficult to undertake and resource intensive. Many NGOs wanted to become much better at learning lessons from previous advocacy in order to improve future work, to satisfy donors on the viability of advocacy activities and to assess the impacts of advocacy on its targets. None of this is easy to measure because it is difficult to distinguish the NGO's own efforts from other influences on social change.

Lack of documentation is also of concern; an Oxfam Australia internal report into advocacy agreed with a range of NGO staff comments that the organisations need to document activities much more thoroughly as a basis for effective evaluations of campaigns. All NGOs agreed that effective evaluation was easier if an appropriate process was built in from the start of a campaign to enable evaluation at regular intervals. Limitations in time and resources, however, often prevent such practices.

Funding NGO advocacy

There is a perception amongst NGOs that many donors are unwilling to support advocacy work financially (Sogge and Zadek 1996: 88). This perception is shared by some of the case study organisations. For example, for WVA, this perception reflected reality, as donations to its Action Network scheme for advocacy accounted for less than 5 per cent of total annual donations to WVA. Similarly, *Angli*CORD knows its supporters largely preferred their donations to go to overseas projects and, therefore, the question of funding advocacy work provides a dilemma for the NGO.

Funding advocacy work was thus a constant challenge for most NGOs. Five NGOs allocated funds specifically for advocacy work; only *Angli*CORD did not. ARHA and FDC do not need to 'find' specific funds for advocacy, as it is their core work. The other four NGOs are in quite a different situation, all raising funds primarily from a combination of donations from the public and grants from AusAID and other organisations. Both INGOs have dedicated policy departments for research and lobbying and solicit funds specifically for advocacy work via non-tax-deductible donation schemes. However, both Oxfam Australia and WVA found that most supporters preferred to support more tangible work or to donate to a tax-deductible fund. Unsurprisingly, for *Angli*CORD, IWDA, Oxfam Australia and WVA, finding adequate funds for advocacy work was an ongoing battle.

Both the faith-based NGOs made a particular effort to educate their supporters on the role of advocacy. *Angli*CORD believed only a minority of its supporters saw the benefit of financing advocacy, and WVA faced the same dilemma when it first moved into advocacy. While WVA believed its supporters now understand the important role of advocacy work, they have been less forthcoming with funds for this. Although *Angli*CORD and WVA represent the smallest and largest of the case study NGOs, both are Christian NGOs with a largely conservative donor base which has a preference for donations to be directed to overseas programmes rather than domestic advocacy. Oxfam Australia, on the other hand, has built a reputation as an activist NGO and, consequently, tends to attract people who want the NGO to be an active advocate. However, even support for Oxfam Australia rarely translates into donors funding advocacy rather than project work.

Many NGOs receive some government funding, although such funding is a contentious issue among activist NGOs (see, for example, Morrow 1997: 4; Smillie 1999b: 19; Hailey 2000: 404; Vaux 2001: 204). Critics warn that, while official funds allow NGOs to expand and build a higher profile, they may also divert NGOs from their activist mission. An NGO's advocacy agenda may be compromised if the NGO starts to rely on funding from official donors and allows donor priorities, rather than communities in the South, to influence strategies. Commins (1997: 141) studied this relationship in some depth in the late 1990s, and argued that 'the growing incorporation of NGOs into the policy agendas of donor agencies' had introduced 'a serious complication' to the relationship between donors and activist NGOs. The challenge for NGOs will be to determine whether the receipt and use of government funds enables them to have greater impact in their operations, whether they will become 'domesticated by their dependence on public sector monies' or whether there are contradictions between receipt of bilateral or multilateral funding and NGOs' 'stated commitment to serving the needs of low income communities' (Commins 1997: 154).

Australian NGOs receive a relatively small amount of development funds from Australia's bilateral aid agency, AusAID, compared to those received from AusAID by the private sector. Whereas ten private companies were awarded contracts with AusAID to deliver aid projects worth $1.2 billion in 2000/01, about 100 Australian NGOs competed for $104.8 million of official aid project money in the same year. By 2003/04 AusAID delivered aid projects worth $1.97 billion but the Australian NGO share had dropped to $94.8 million (AusAID 2005a, 2005b). FDC competes with private companies for AusAID contracts as it is not eligible for AusAID accreditation and cannot, therefore, compete for NGO funds.

Four of the case study NGOs had received project funding from AusAID. FDC was not eligible and ARHA had not received AusAID funding. Current AusAID accreditation criteria require an NGO to be a registered charitable or benevolent institution which implements aid projects in developing countries; hence ARHA did not qualify. Among the other four NGOs, the ratio of funds received from AusAID to funds raised from the public varied considerably both between NGOs and also annually within the NGO. For example, in the financial year 2001/02, Oxfam Australia derived over 66 per cent of its income from supporters and only 12.7 per cent from AusAID (OCAA 2002b: 20). In 2003/04, this had risen to 76 per cent from supporters and 17.6 per cent from AusAID (OCAA 2004: 23). Oxfam Australia was mindful that receipt of too much funding from AusAID could compromise its wish to remain largely an activist NGO.

For IWDA, in the mid-1990s, AusAID funds were approximately ten times the amount of donations from supporters. By 2001, however, the amounts were roughly equal and, by 2002, supporter donations were double that from AusAID. IWDA did not, as such, make a decision to seek less funding from AusAID; rather, AusAID funding became more difficult for small NGOs to acquire. Fortunately for IWDA, supporter donations more than doubled at that time. By 2004, AusAID funds again played a more significant role in IWDA's budget, accounting for 54 per cent of funding, with supporter donations, while continuing to increase in real terms, accounting for 36 per cent (IWDA 2004: 9).

*Angli*CORD has always raised significantly more from its supporters than it has received from AusAID. In the financial year 2000/01, 68 per cent of its funds came from supporters whereas 29 per cent came from AusAID. By 2004/05, *Angli*CORD supporter donations were three times the 2003/04 figure and AusAID funds had doubled, effectively trebling *Angli*CORD's funding over the year. By mid-2005, supporters' donations accounted for 82 per cent of *Angli*-CORD's budget while AusAID funds represented only 16 per cent of funding, even though the actual amount had doubled over the year (*Angli*CORD 2005: 7).

As a 'super-NGO', able to commit the resources and expertise necessary to attract large amounts of AusAID money, WVA saw grants from AusAID significantly increase as a percentage of income over the five years until 2001. In 1997, AusAID funds accounted for around 13 per cent of total income for WVA whereas in 2001, even though supporter income was nearly 50 per cent higher than 1997 figures, AusAID funding had increased to approximately 27 per cent of total income. This percentage dropped in 2002 as public donations continued to increase, and AusAID funding accounted for approximately 22 per cent of total WVA income. However, the trend had reversed by 2003 and AusAID funds as a percentage of WVA's total funding had dropped to below 10 per cent. By 2004, supporter donations had continued to increase as a percentage of what is, by far, Australia's largest NGO budget, and the proportion of AusAID funding within that budget had dropped to 7.5 per cent (World Vision Australia 2004: 8).

AusAID funding through contracts contributed about 4 per cent of FDC's total income in 2000. In the following year it accounted for almost 9 per cent of total income and, in the financial year 2001/02, AusAID contracts accounted for almost 50 per cent of FDC's total income. FDC's investment income decreased by a quarter during 2003, but the increase in funds from AusAID was, nevertheless, substantial in real terms.

Although the extent of dependence on government funding has varied over time for some of the NGOs, most have gained some income from AusAID. Although grounds for concern about the relationship between official donors and activist NGOs exist (Hulme and Edwards 1997: 275), all the NGOs believed official funding did not compromise the extent and type of their advocacy, although all are well aware that this could happen if NGOs lose sight of their activist mission in favour of funding possibilities. Oxfam Australia was partic-ularly determined to never compromise its activist intentions by allowing funding considerations to influence advocacy work, with one staff member asserting: 'Oxfam Australia would give up all funding before we would move away from our social activist role . . . [as] after all, that is who we are.' While WVA does not consider itself to be an activist organisation, it is committed to advocacy and would not allow funding considerations to prevent this. *Angli*CORD and IWDA shared similar views.

AusAID has publicly said that it 'fully supports NGO advocacy work' (AusAID 1995: 6). However, Glanznig (1996: 133) argued that AusAID only supports advocacy if not directed against Australian government policies. AusAID has itself argued that, in the developing world, through advocacy and lobbying work, NGOs are playing a significant role in 'demanding public accountability from government and business sectors' as well as promoting pluralism and protecting the interests of minority groups and others considered disenfranchised by mainstream political processes (1995: 6). In contrast to this, many of the 65 surveyed NGOs commented that AusAID was not at all supportive of NGO advocacy when NGOs made public statements against AusAID policies, the minister or the government. However, apart from the occasional conversation where a staff member of an NGO felt 'warned' by an AusAID member, no evidence was presented of tangible action by AusAID against any NGO which had opposed government policies.

It is not possible to use AusAID funds for advocacy as funds are strictly monitored by the agency. Before an NGO can apply for AusAID funds it must be 'accredited'. During the accreditation process, an assessment is made about the ability of an NGO to work cooperatively with AusAID,[6] and AusAID's own documentation indicates that AusAID does not fund NGO advocacy.[7] Some NGOs believe that AusAID funding is denied to those NGOs which join advocacy campaigns directed against Australian government policies, but none had direct evidence for this. One had personally 'felt the heat very much from the minister' over some recent advocacy against a government policy, with the NGO being asked to provide a 'letter of comfort' to assure AusAID that no government funds were used in that advocacy. The NGO believed that, even on that occasion, AusAID's and the minister's displeasure did not translate into funding denial.

Another NGO reported that, on one occasion several years ago, in response to a press release regarding Australia's aid budget, the response from AusAID was 'a bit over the top'; this was the only occasion the NGO had felt 'real heat' from AusAID. When funding was subsequently denied for a project which the NGO believed would have been accepted, AusAID's response 'caused an element of doubt', but the NGO was prepared to accept that perhaps its proposal was not 'up to scratch'.

Several NGOs said they knew of instances where 'payback' had occurred. One reported occasions when NGO staff tried to be honest with AusAID and had doors closed to them as a result. Another commented that, in the NGO community, it is widely accepted that criticism of AusAID gives an NGO or a staff member a reputation within AusAID for trouble-making. Specific occasions were cited when an NGO staff member had been told something like 'I wouldn't do this if I were you'. Although it is not possible to prove that funds were withheld by AusAID because of advocacy work, the perception was that this had occurred. Another NGO spoke of an instance when an ACFID campaign criticising AusAID led to 'harsh treatment' by AusAID of one or more of the participant NGOs. An NGO staff member said their organisation had worked hard for some time to develop a cooperative working relationship with AusAID and the NGO had never been denied funding in spite of participation in campaigns against AusAID policies.

Another NGO staff member 'absolutely believes' that 'NGOs that are funded by AusAID *do* [interviewee's emphasis] feel constrained – rightly or wrongly – in criticising AusAID' and that the current Liberal–National Party coalition government 'seems to feel that any strong advocacy position taken by any agency in any field which does not support current government policy is somehow reprehensible'. This was contrasted with previous governments, both Liberal–National Party coalition and Labor, which were perceived to have understood that 'there is a role for constructive criticism and advocacy which can, in fact, be helpful in the formulation of policy'.

Commins (1997: 146) argued that funding has not been denied to WVA as a result of any criticism made about the Australian aid programmes or government policies, even though WVA has experienced, and witnessed, 'direct and sometimes quite strong exchanges between the government and NGOs on various matters'. WVA policy was that, in order to keep an open relationship with the government, it advised AusAID of any issues which might be critical of the government before making public statements. Moreover, WVA developed contacts with like-minded staff within AusAID and, if appropriate, provided public comments to strengthen the negotiating position of those within AusAID who were supportive of specific reforms.

The views of AusAID staff about the role of NGO advocacy vary considerably and some are more willing to listen to advocacy than others. However, at least one NGO believed that most AusAID staff 'do not recognise the important space that NGOs occupy as part of civil society'. High staff turnover at AusAID also caused problems in continuity for NGOs, impeding any real understanding of the NGO advocacy role, and many new staff had little or no knowledge of aid, NGOs or advocacy. However, high staff turnover could be positive; if the NGO was 'off-side' with an AusAID staff member, the chance existed to rebuild the relationship with a new staff member. At the time of the research, AusAID was addressing issues caused by high staff turnover, and new AusAID staff attended a compulsory formal induction course, which included information about the role that NGOs play in development as well as training about human rights. Additionally, AusAID staff can attend more specific courses, which some NGOs run from time to time, on topics such as NGO accountability and programme monitoring.

Two senior AusAID staff with extensive experience of the AusAID/NGO relationship unsurprisingly said that AusAID 'absolutely' understands and supports the NGO advocacy role. Both rejected suggestions that AusAID may understand but not support the advocacy role or may discriminate against NGOs which advocated against AusAID policies; both said AusAID has systems in place to prevent any discrimination. Funding is allocated by a set formula, the process is fair and transparent and the selection process involves a joint NGO/AusAID selection panel. AusAID staff observed that denial of funding was often because an NGO application was 'just not good enough'. In such circumstances, NGOs may blame past advocacy. As a bilateral agency, AusAID *must* reflect current government programmes and this is not discretionary. Therefore, NGOs must be prepared to also reflect those programmes to attract funding. Necessarily, there will be 'ongoing tension between AusAID and NGOs that advocate development strategies with different theoretical bases to the models of development being advocated by AusAID' (Glanznig 1996: 133); thus, negotiating a path between staying true to their activist mission and attracting official funding remains a challenge for many NGOs.

Themes in Australian NGO advocacy

In the 1970s, NGOs mostly lobbied their governments about the quality of government aid, international power politics, and the role that the socio-economic systems in the aid-giving countries play in hindering development in the developing countries (Lissner 1977: 224). These themes continue to provide the basis upon which much NGO advocacy work is built. Other specific themes which were the subject of advocacy work by the NGOs have included:

- Aboriginal Australia
- asylum/refugees
- capital flows/Tobin tax
- child soldiers
- core labour standards/child labour
- corporate responsibility
- corruption
- debt relief
- globalisation
- health and education
- HIV/AIDS
- human trafficking
- infant formula
- landmines
- microfinance
- poverty
- protecting the environment
- reform of international institutions
- reform of international trade rules
- small arms

Figure 3.2 Oxfam's Make Trade Fair campaign: protestors at the Sydney Opera House, 2002

- street children
- violence.

In addition to advocating about the general interests of a development NGO, for example aid, debt relief and poverty, NGOs choose additional themes that reflect their particular interests. For example, *Angli*CORD has advocated about aid, refugees, debt relief and HIV/AIDS. The first three reflect the interests of most aid NGOs, and the latter *Angli*CORD's particular interest in Africa. In the case of IWDA, in addition to aid, refugees, debt relief and poverty, the NGO advocated about issues IWDA believes impact on women, including Aboriginal Australia, corporate responsibility, the environment, reforming of international institutions, violence and trafficking. Similarly, in addition to aid, refugees and debt relief, ARHA's specialisation in reproductive health issues is reflected in advocacy about health, education and HIV/AIDS.

Since the start of the 1990s, WVA advocated about more issues than any other case study NGO. This was, in part, a reflection of its size, but also of WVA's early advocacy philosophy to join campaigns rather than run its own. Even after WVA became experienced in campaigning, it continued to enjoy the buffer from criticism that working collaboratively lends NGOs. Towards the end of the 1990s, WVA did target some specific areas of interest, for example child rights and HIV/AIDS, in running its own campaigns.

Future advocacy, according to Australian NGO staff, is likely to follow international trends in focusing on key issues such as the protection of human rights, promotion of civil society and good governance, the debt burden of poor countries, women in development, the protection of the environment, and equity and other concerns related to globalisation (Redden 1999: 5; Van Rooy 2000: 309; Ollif 2003: 74). However, all stressed that, as advocates, organisations must be ready to identify and react to new needs. As well, each NGO tends to have its own priorities for advocacy to continue to guide it. For smaller NGOs, future choices about advocacy will be based largely on issues ACFID takes up. In the case of the INGOs, as well as continuing to be active in ACFID campaigns, both Oxfam Australia and WVA take direction in their advocacy from Oxfam International and World Vision International respectively.

Future challenges for advocacy

The issue of foremost concern to NGOs engaging in advocacy is that government conservatism presents a real challenge for future advocacy in Australia. In 2001, Oxfam Australia's (then) executive director publicly expressed concern that, in Australia and elsewhere, people's right to speak out was under attack by conservative forces in government (Hobbs 2001: 4). According to the (then) ARHA chief executive, although conservative disinterest is the greatest challenge facing Australian advocacy NGOs, they must not allow fear of loss of government funding to impede their advocacy. After all, in essence advocacy involves challenging government policies and, in continuing to advocate about issues of inequality and social justice, NGOs must be braver and more outspoken – at both national and international levels (Proctor 2000b: 6).

Further challenges related to NGO advocacy include maintaining accountability in the face of increased public exposure to, and commensurate public scrutiny of, NGO activities. In increasing their role in shaping public opinion, NGOs will be expected to be accountable to a range of stakeholders including their partners, supporters, donors and other institutions. Secondly, NGOs face the need to attract adequate funding, mentioned by small and 'super' NGOs alike. It is difficult for an NGO to effectively demonstrate the benefits of advocacy to donors and, therefore, it can be difficult to increase revenue for advocacy work. The faith-based NGOs find encouraging their supporters to fund advocacy a particular challenge, with supporters preferring all donations to fund project work overseas. To address this dilemma, Roche (1999: 192–233) argued the importance of NGOs demonstrating that advocacy both is cost-effective and makes a positive difference to people's lives. NGOs are aware, also, that they must be careful not to be co-opted to the agenda of official donors. As well, the challenge to become much better at articulating the NGO case, according to one NGO, requires avoiding petty comments and the emotional or high moral ground, which jeopardise the success of advocacy. Building more deliberate alliances for advocacy purposes is also important. All case study

NGOs intended to continue to seek out appropriate alliances for advocacy, with most believing strong alliances are vital for their success and will become more so. A member of ACFID's advocacy department also commented that the challenge for NGO advocacy is to reach the non-converted and convertible. Advocacy often preaches to the converted and, therefore, it is imperative for NGOs to find ways of being heard by the wider public; NGOs have still to learn how to fully exploit the media. Finally, the challenge for the INGOs is to move beyond present aid policy work into 'the wider, more difficult and more contentious arena of how to reshape the forces that are presently driving the global economy' (Commins 1997: 155).

Conclusion

Like their global counterparts over the last one to two decades, many Australian NGOs have adopted advocacy as the strategy most likely to achieve significant reductions in levels of global poverty. However, lack of adequate resources is the single most important impediment to Australian NGO advocacy. Only the smallest NGOs have not engaged in advocacy, or only do a limited amount, with any advocacy primarily within an ACFID or 'super-NGO' alliance. Australian NGOs that are members of a 'super-NGO' global alliance have emerged as the most active as advocates in Australia. However, most NGOs still have to convince their donor base of the need to adequately fund advocacy as an essential component of their work.

If they are to combat the lack of resources and increase the impact of their advocacy, the importance of alliances to NGOs is clear. Those Australian NGOs engaged in advocacy are eager to cooperate with their counterparts to achieve positive outcomes. ACFID has a very important role in facilitating this cooperation and providing advocacy support to large and small NGOs alike. Both ACFID and the 'super-NGOs' play an important role in fostering alliances when smaller NGOs venture into advocacy work; these have allowed smaller NGOs to contribute their own special expertise to campaigns largely resourced by other NGOs.

Ultimately, it is difficult for NGOs to demonstrate a direct connection between advocacy actions and positive social change (Welch 2001: 272). For NGO staff, assessing the exact impact of their own efforts when advocacy appears to be successful is problematic because of the many forces involved. However, a number of elements were identified as important for the success of advocacy. Firstly, research must be thorough, scrupulously accurate, well documented and backed up with demonstrated public support for the cause. Secondly, alliances can bring a diversity of resources and expertise to campaigns, as well as providing clout through numbers. Thirdly, appropriate use of the media is important. The media can make the NGOs' concerns heard, thus setting the agenda. Through the media, an NGO demonstrates to its supporters that it is being taken seriously, and the media can also be used to rally the support of like-minded people. Fourthly, communications technology has become critical to effective advocacy and is an important way to increase supporter participation and disseminate information about advocacy work to a national and/or global audience. Like NGOs worldwide, Australian NGOs will increasingly use lobbying and advocacy in an effort to improve or change policies of governments and global institutions in attempting to increase standards of living and life choices for those suffering the ravages of poverty and disadvantage.

Part II

Towards global equity?

Internationalisation, Oxfam and the World Bank

4   Global action

International NGOs and advocacy

Ian Anderson

Northern NGO advocacy has come a long way since the early 1970s' campaigns which Clark describes as 'poorly financed and run by highly committed but inexperienced volunteers but [which were] highly effective at capturing the public imagination' (in Edwards and Hulme 1992: 197–8). NGO advocacy has become more focused and more strategic and has made more effective use of the media. NGOs have learnt to gain access to and use the political processes, structures and institutions of their home countries as well as those of the multilateral agencies. This evolution of NGO advocacy has led to more effective interaction between NGOs and official agencies, to alliances between Northern and Southern NGOs as those from the South have expanded their advocacy into the international arena, and to alliances between broad-based development and relief NGOs and specialised campaigning groups and networks, including environmental organisations.

Despite this progress, there are widely divergent views about the merits of NGO advocacy. Dating from the early 1990s, these views range from being highly positive to being deeply sceptical. Both Clark (1991) and the UNDP (1993) attributed significant (but largely unsupported) positive results and strengths to Northern NGO advocacy, Sogge (1996a) discerned a lack of any real knowledge of its effects, and Korten (1990) wrote disparagingly of NGOs' lack of vision. Edwards (1993), who has written extensively on Northern NGO advocacy, identified four key strategic weaknesses in the international advocacy of British NGOs as representative of Northern NGOs: absence of a clear advocacy strategy; the failure of NGOs to build strong alliances to broaden and so strengthen their advocacy voices; failure to develop credible alternatives to established orthodoxies, which, critics suggested, required more research by NGOs and a more conscious linkage of NGO field experience with the development models they adopt; and relations with official donors which NGOs are afraid to criticise, while being heavily reliant on their funding. Additionally, Northern NGOs' role as legitimate advocates for the Southern poor came under scrutiny as Northern NGO advocacy evolved and Southern NGOs themselves became increasingly involved in advocacy beyond their national borders. Northern NGOs are being challenged on issues which include the changing nature of relationships between Northern and Southern NGOs and calls for new forms of alliance between them, Southern expectations of their Northern counterparts, and tensions concerning who should determine the development agenda. More recently, NGOs have been challenged in relation to the very nature of their roles as agents in the resource-transfer paradigm which has historically characterised Northern and international NGOs. The challenge to NGOs has been described in the following terms:

> Moving from development-as-delivery to development-as-leverage is the fundamental change that characterises this shift and it has major implications for the ways in which NGOs organise themselves, raise and spend their resources, and relate to others . . .
>
> The fundamental question facing all NGOs is how to move from their current position – as unhappy agents of a foreign aid system in decline – to where they want to be – as vehicles for international cooperation in the emerging global arena.
>
> (Edwards, Hulme and Wallace 1999: 30, 134)

Other commentators have identified further weaknesses, including NGOs' inability to demonstrate, through evaluation, the effectiveness and impact of their advocacy (Tendler 1982; Sen 1987; Clark 1991; Edwards and Hulme 1992, 1995; Fowler 1995; Fowler and Biekart 1996; Saxby 1996; Sogge 1996a; Roche 1999; Hudson 2000b, 2001b; Davies 2001; Coates and David 2002; Kelly 2002).

It is broadly accepted that structural macro-reforms are essential if the fundamental causes of poverty are to be redressed. Watkins summarises the need for such reforms as 'requiring a transformation in attitudes, policies and institutions' and 'a fundamental redirection of policy on the part of other foci of power including the UN, international financial and trade organisations, corporations (TNCs), official aid donors and NGOs' (1995: 216, 217). This is one of the major challenges facing Northern NGOs in their advocacy: how, by employing strategies which maximise their effectiveness and impact, they will be able to 'address the structural causes of poverty and related injustice' (OI 1999a: 4).

Although much has been written about NGO advocacy as an increasingly prominent aspect of their work, little is known about its efficacy. In adding to the existing body of knowledge about the extent and effectiveness of the advocacy of Northern and international NGOs, two avenues of research are detailed here. The first addresses the generalised criticisms of Northern NGO advocacy and assesses the validity of these. The second explores the workings of advocacy and adds to existing knowledge about the outcomes of Northern or international NGO advocacy.

Empirical data are required to assess generalised criticisms of Northern NGO advocacy. A survey was designed to collect data relevant to the oft-cited criticisms of Northern NGO advocacy, following a standard path in social science principles (Burns 1998; Frankfort-Nachmias and Nachmias

2000; Anderson 2003). This chapter reports the findings of the survey. Then, preparatory to the following chapter, which evaluates the policy outcome effects of Oxfam International's World Bank-oriented advocacy, it details how members of Oxfam International (OI) structured their relationships to increase their separate and collective advocacy capacity.

Northern NGO advocacy

Data provided by responding NGOs enabled an assessment of several issues: NGO incomes, including the extent of official donor funding; the proportion of spending on development projects, humanitarian relief and advocacy in its several forms; how advocacy programmes are staffed; the principal issues about which NGOs have advocated; whether advocacy is regarded as a strategic component for the achievement of NGOs' mission; how advocacy is evaluated and how evaluation results are shared with stakeholders; and what advocacy alliances are maintained. To shed light on some of these issues, surveys were sent to 56 larger Northern NGOs. The sample was drawn from the major Northern NGO networks in Europe, North America and the developed Asia-Pacific economies. Data provided by the then 11 affiliates of OI augmented the sample.

By covering the period from the early 1980s, the survey sought to discern trends in Northern NGO advocacy over an extended period. This was intended to reveal whether the increasing prominence given to advocacy in the literature since the early 1980s was matched by actual Northern NGO resource allocation. If so, this would suggest that NGOs have been heeding calls for greater emphasis on advocacy within their poverty reduction mission, and that NGOs themselves have discovered this to be the case. At the time of conducting the initial survey in early 1998, 1996 was the latest year for which data were available (Anderson 2003). To obtain insights into developments since 1996, each of the NGOs which provided data was requested to respond to a set of further questions in 2005. Fourteen of the 23 NGOs which provided data to 1996 responded to the further questionnaire.

NGOs based in OECD member states were selected as representative of Northern NGOs based in mature and developed economies, with the OECD directory used to select NGOs based in OECD member states (OECD 1992). Two factors were considered as selection criteria in the survey: budgeted expenditure as indicative of the scale of an organisation's size and operations, and participation in development education as one of the streams of NGO advocacy (the OECD directory did not list advocacy as a separate entry). All NGOs with budgeted 1990 expenditure not less than that of Oxfam Canada ($US7 million) were included in the survey, except for those which were primarily involved in environmental rather than human development programmes. Oxfam Canada's budget was used as the benchmark because it was known to be engaged in a full range of development NGO activity, including advocacy, and was therefore comparable with other OECD country-based development NGOs.

Table 4.1 Total number of NGOs surveyed and response rates

	International/ Northern NGOs		Oxfam International affiliates		Total	
	N	% of NGOs surveyed	N	% of NGOs surveyed	N	% of NGOs surveyed
NGOs surveyed	56	100	11	100	67	100
NGOs from which a response was received	41	73.2	11	100	52	77.6
NGOs which provided survey data	12	21.4	11	100	23	34.3

Of the 56 NGOs which were surveyed, 78 per cent responded; 34 per cent of the total number surveyed actually provided data. Some responding NGOs were unable to provide data owing to time constraints and others advised that they did not engage in advocacy. The development projects and humanitarian relief expenditures of the NGOs which provided data represented 13.6 per cent of total grants by OECD member countries' NGOs in 1996. The number of NGOs surveyed and the responses received are summarised in Table 4.1.

Survey data are thus biased towards the then 11 members of Oxfam International. Those others which provided data included three from one international NGO network and two from another.[1] The survey data, analysis and conclusions therefore reflect these limitations.[2]

Income from government sources and advocacy expenditures

The increasing proportion and scale of NGO funding from official donors, in the form of governments, government organisations, regional groupings of governments and multilateral agencies, are a feature of Northern NGO revenues highlighted in the literature. This is perceived to be constraining and 'emasculate[s] NGO attempts to serve as catalysts and advocates for the poor' (Edwards and Hulme 1992: 20; see also Minear 1987; Arnold 1988; Korten 1990; Smillie and Helmich 1993; Roche 1994; Smillie 1995; Sogge 1996a). Therefore, data about official funding, the proportions of NGO spending on advocacy, and the extent to which participating NGOs are active in advocacy were sought and compared.

The survey also requested (where available) information about allocation of expenditures between the four principal streams of advocacy as defined (development education, campaigns, lobbying and supporting research), in order to discern, insofar as expenditure is a measure, the relative importance NGOs have attributed to particular forms of advocacy. Overall, the goal was to test Clark's hypothesis that, despite the broadly accepted view that advocacy is the strategy most likely to influence significant poverty reduction, NGOs have directed few resources to advocacy (1991: 147).

It was not possible to determine a correlation between proportions of government funding received by NGOs and advocacy expenditures; there is no consistent link between the two factors and, particularly in relation to the non-Oxfam International/Northern NGOs, too many cases where insufficient data preclude reliable analysis.

Because of the Oxfams' largely independent autonomous operations prior to 1995, it was impossible to analyse them as a homogeneous group, so Oxfam International affiliates are reviewed separately. The lack of correlation between levels of government funding and advocacy expenditures is, in the case of the Oxfams, most apparent in Table 4.2 in relation to Oxfam America and Intermon.[3] Oxfam America, which as a matter of policy accepts no government

Table 4.2 Oxfam International affiliates: funding from governments/government bodies and advocacy expenditures, 1984[1] and 1996 (percentage of total)

Respondent	Government funding			Advocacy expenditures		
	1984	*1996*	*Change – compound % p.a.*[2]	*1984*	*1996*	*Change – compound % p.a.*[2]
1 Community Aid Abroad (Australia)[3]	56.8	39.3	10.6	0.1	8.9	40.0
2 Intermon (Spain)	2.8[b]	52.3	80.4	0.0[b]	11.0	Insufficient data
3 Novib (Netherlands)	64.1[a]	70.7	5.9	5.4[a]	4.0	(3.0)
4 Oxfam America	0.0	0.0	0.0	10.4	5.3	1.1
5 Oxfam-in-Belgium	57.8	79.2	16.0	N/A	5.7	Insufficient data
6 Oxfam Canada	62.5	51.2	5.2	5.5	5.9	7.3
7 Oxfam Great Britain	9.2	33.1	24.2	4.4	6.7	16.6
8 Oxfam Hong Kong	18.2[a]	12.6	6.9	3.2[a]	5.4	46.2
9 Oxfam Ireland[4]						
10 Oxfam New Zealand	33.3[a]	46.1	31.6	2.6[a]	3.0	18.9
11 Oxfam Quebec	23.9	58.3	19.7	2.3	1.2	5.9

Notes:
1 Or earliest year for which data are available: a – 1992; b – 1986.
2 Calculated by reference to change in amount (expressed in currency of response unadjusted for price movements) of government funding or advocacy expenditures from 1984 (or first available year) to 1996.
3 Name changed to Oxfam Australia in 2005.
4 Since the survey was undertaken, Oxfam Ireland was formed by division of Oxfam United Kingdom and Ireland, creating it and Oxfam United Kingdom.

funding, decreased its advocacy expenditures as a proportion of total expenditures from 10.4 per cent in 1984 to 5.3 per cent in 1996. Over this period, advocacy expenditures (unadjusted for price movements) increased by 1.1 per cent per annum compound. In contrast, Intermon, whose government funding increased from 2.8 per cent of total expenditures in 1986 to 52.3 per cent in 1996 (a compound growth rate of 80.4 per cent per annum), had by 1996 increased its advocacy expenditures from zero in 1986 to 11 per cent of total expenditures by 1996. This was the highest proportion of expenditure on advocacy of all Oxfam International affiliates. Other than for Oxfam New Zealand in its start-up phase, the next greatest rate of increase in government funding was experienced by Oxfam Great Britain at 24.2 per cent per annum. Although that rate of growth in government funding exceeded the 16.6 per cent of growth in advocacy expenditures, Oxfam Great Britain's expenditures on advocacy increased from 4.4 per cent to 6.7 per cent of total expenditures. The lack of correlation between government funding and advocacy expenditures is further highlighted by the comparison between Oxfam Canada and Oxfam Quebec, both of which received in excess of 50 per cent of total income from government and both of which, being based in Canada, might have been expected to be subject to similar government influences. Oxfam Canada's advocacy expenditures have remained consistently above 5 per cent of total expenditures and have grown at a rate higher than those for Oxfam Quebec, where advocacy expenditures have declined as a proportion of total expenditures from 2.3 per cent in 1984 to 1.2 per cent in 1996. This comparison between the two Canadian-based Oxfams suggests a difference in orientation rather than any government influence on advocacy.

There was a marked difference in levels of advocacy expenditure for the non-Oxfam International/Northern NGOs (Table 4.3). One of the highest levels of expenditures on advocacy at 18.0 per cent of total expenditures was I.2, which at 14.8 per cent in 1996 received near the lowest level of government funding of the responding NGOs. Conversely, those receiving the highest levels of government funding (ranging from 63.1 per cent (I.9) to 89.8 per cent (I.5) in 1996) spent little on advocacy; I.5's 4.7 per cent 1996 advocacy expenditures (down from 10 per cent in 1984) would suggest some, although declining, priority being attached to advocacy. Although the data suggest some correlation between advocacy expenditures and government funding, the sample was too small to confidently draw any real conclusion.

An overwhelmingly greater proportion of NGO spending continued to be directed towards the more traditional forms of NGO assistance in the form of development projects and humanitarian relief (Table 4.4), suggesting that Clark's hypothesis (1991) that few NGOs have committed significant resources to advocacy is valid.

NGOs are constrained in their allocation of resources by a combination of private and official donor funding pressures and expectations, by limitations on NGO 'political' activity and by their inability to demonstrate, through evaluation and the publication of evaluation results, the effectiveness of their advocacy.

Table 4.3 International/Northern NGOs: funding from governments/government bodies and advocacy expenditures, 1984[1] and 1996 (percentage of total)

Respondent		Government funding			Advocacy expenditures		
		1984	*1996*	*Change – compound % p.a.*[2]	*1984*	*1996*	*Change – compound % p.a.*[2]
I.1	ActionAid	N/A	13.0	Insufficient data	N/A	6.5	Insufficient data
I.2[3]		20.0	14.8	(2.9)	13.0	18.0	Insufficient data
I.3	Bread fuer die Welt	N/A	0.0	Insufficient data	N/A	N/A	Insufficient data
I.4		35.5[a]	77.2	30.2	Neg.[b]	Neg.	38.4
I.5		77.2[b]	89.8	20.6	10.0	4.7	(1.4)
I.6		22.4[a]	34.5	30.3	4.7	0.0	Insufficient data
I.7	No data	N/A	N/A	N/A			
I.8	Lutherhjalpen Church of Sweden Aid	21.2	27.5	6.9	5.0	8.0	Insufficient data
I.9		59.1[b]	63.1	17.7	0.0	0.0	0.0
I.10	Trocaire	3.4	45.0	25.0	10.3	7.5	3.0
I.11		5.4	14.8	21.8	0.1	1.6	18.3
I.12		13.8	12.7	8.8	3.8	5.6	15.3

Notes:
1 Or earliest year for which data is available: a – 1988; b – 1992.
2 Calculated by reference to change in amount (expressed in currency used in response unadjusted for price movements) of government funding or advocacy expenditures from 1984 (or first available year) to 1996.
3 NGOs have only been named where they gave the author permission to do so. Where NGOs have restricted the attribution or publication of data, they are numbered.

Table 4.4 NGO spending on advocacy as a proportion of total spending, 1996 ($US million)

	International/ Northern NGOs	*Oxfam International affiliates*	*Total*
Total expenditures	576.1	370.8	946.9
Total advocacy expenditures	15.8	23.0	38.8
Advocacy expenditures as percentage of total expenditures	2.7	6.2	4.1

Although constraints on resource allocation may exist, survey data did not indicate a direct or uniform correlation between government funding and NGO advocacy expenditures.

That the surveyed NGOs applied just 4.1 per cent of total expenditures to advocacy suggested limited confidence in it as the strategy most likely to contribute to their expressed poverty eradication mission, or in NGOs' ability to convince their constituencies of its importance (even allowing for the constraints noted above and Northern NGOs' claim that their legitimacy for advocacy is derived largely from their development and humanitarian relief experience).[4] This allocation of resources into NGO advocacy may be contrasted, for example, with that of the environmental NGO Greenpeace, which embraces an action-oriented strategy, which exists as a 'catalyst for change' and which has demonstrated the ability to mobilise large numbers of people in pursuit of specific achievable objectives (Greenpeace International 2006: 1, 3). Greenpeace therefore embraces a wholly advocacy-focused strategy, compared to that of the development NGOs, whose levels of allocation of resources to advocacy confirm that, in the case of Northern development NGOs, 'Advocacy may be seen as *important* but it is not *urgent*. Consequently it is easily squeezed out by the day-to-day dilemmas and crises arising from the project activities, from donor pressures and from media enquiries' (Clark 1991: 147). Remarkably, up until 1996, little had changed in the allocation of expenditure for over a decade.

Advocacy staffing and advocacy strategy

As a further indication of the resources allocated to advocacy, and to assess whether the failure to integrate it into the fabric of NGOs represents a related key strategic weakness (Edwards 1993: 168), the survey sought data about staffing. Advocacy staffing may also be linked to whether NGOs have a stated policy of advocacy as a strategy for use in pursuit of their objectives. The number of senior and middle management staff committed to advocacy indicate the level of priority attached to it, its integration into the fabric of the organisation and the level of engagement with governments, multilateral agencies and the media sought by NGOs. Thus, the greater the senior and middle management staff time engaged in advocacy-related activity, the greater is the assumed organisational and strategic importance attached to advocacy.

Among the responding non-Oxfam NGOs, 10 of 12 advised they had a stated advocacy policy, while for the Oxfam International affiliates the numbers were 7 of 11. This generally correlated with allocation of staff time and the dedication of specialist staff to advocacy. Of the non-Oxfams, only one (I.9) with an advocacy policy reported having no specialist staff, while among the Oxfams only Oxfam Hong Kong was in this position. Conversely, one non-Oxfam (I.5) and one Oxfam (Belgium) indicated having specialist advocacy staff but no advocacy policy. Of the responding non-Oxfams, six had senior management staff with advocacy responsibility, while three did not. Data were not available for three. For the Oxfams, only Oxfam Hong Kong did not have senior management with responsibility for advocacy, although in three other cases advocacy was only part of the responsibility of senior management. In every case, total advocacy staff numbers in 1996 exceeded or were not less than the average of reported advocacy staff numbers over the survey period. Therefore, over the 12-year survey period, total staff resources for the NGOs' advocacy increased. This is consistent with the data in Tables 4.2 and 4.3, which, where available, reveal generally higher, although limited, levels of advocacy expenditures in 1996 compared to 1984. By 1996, NGOs generally were attaching greater organisational importance to advocacy. The assignment of specialist staff, and generally greater senior management staff engagement, combined with reported expenditures, indicate that advocacy was more integrated into the fabric of NGOs.

A summarised description of the rationale, objectives and issue selection criteria for NGO advocacy was sought to assess, however superficially, the validity of criticism of the slowness of NGOs to embrace, clarify and integrate advocacy as a strategy for achievement of their stated aims (Edwards 1993: 168).

Most of the respondents with an advocacy strategy provided a brief commentary on the rationale, objectives and issue selection criteria for their advocacy. Predictably, responses referred to influencing decision makers and public opinion to bring about change to the benefit of poor people as a principal reason for, and objective of, NGO advocacy. In selecting issues for advocacy, a number linked them to field experience, to their assessment of the prospects of actually contributing to positive change and to influencing opinion within their home country constituencies. However, despite the recurrent references to overseas field experience influencing topics for advocacy, only two Oxfam International affiliates and no non-Oxfams referred to consultation with Southern NGO partner organisations. In other words, 21 of 23 respondents did not consult their Southern 'partners'. Unless Northern and international NGOs' advocacy is strongly grounded in their field experience, there may be merit in questioning their legitimacy to speak as advocates for the Southern poor, and criticisms of Northern NGOs' failure to build effective partnership relationships with Southern NGOs may be valid (Salmen and Eaves 1991; Roche 1994; Edwards and Hulme 1995).

Advocacy alliances

The failure to build alliances which strengthen and broaden NGO advocacy messages is one of the four principal strategic weaknesses identified by Edwards (1993: 172). Smillie (1995) was critical of NGOs for allowing competition between NGOs rather than coordination and for their failure to follow the lead of the women's and environmental movements, which spread their influence through networks and coalitions of groups following a broadly similar agenda. Advocacy relationships between Northern and Southern NGOs are, however, increasingly seen as essential to the legitimacy of Northern NGOs as advocates

Table 4.5   Advocacy alliances: number of respondents operating advocacy alliances

Category of organisational alliance	International/ Northern NGOs (12 respondents)	Oxfam International affiliates (11 respondents)
Home country	10	11
Regional	9	9
Northern umbrella bodies	10	8
Southern organisations	7	7
Own international network	9	10

who speak for the Southern poor (ICVA 1990; Clark 1991; Edwards and Hulme 1992, 1995; Smillie 1995; Sogge 1996a).

All the NGOs cooperated in some form of advocacy alliance with other Northern umbrella alliances within their own international networks, and with Southern organisations (Table 4.5). It was, though, beyond the limits of the survey to establish the breadth, strength, levels of commitment and effectiveness of those alliances.

Only one non-Oxfam did not participate in advocacy alliances with either Southern organisations or with its own international networks. In the case of the Oxfams, one in each of the 'Regional organisations' and 'Other Northern alliances/umbrella bodies' categories answered in the negative, and two did not undertake advocacy in alliance with Southern organisations. Otherwise, all respondents maintained advocacy alliances in each of the five categories. Two dominant reasons were given for maintaining advocacy alliances: the sharing of experience, expertise and resources; and greater impact and efficiency.

Principal issues for advocacy

Trends in the focus of advocacy over the survey period (measured in four-year periods) and any commonality in the focus of NGO advocacy were examined. For the purposes of the survey, the definition of advocacy included development education, campaigns, lobbying and related research, so responses did not distinguish between types of advocacy activity. The growth in the number of NGO advocacy topics over the survey period, the very wide range of topics and, in the periods 1989–92 and 1993–96, the emergence of several issues around which NGO advocacy coalesced were notable. In the period 1989–92, respondents most frequently reported advocacy in relation to their home country and Northern governments' official aid policy, and the proportion of NGOs advising advocacy on this issue was relatively high. Almost all engaged in debt, landmine and trade-related advocacy over the period 1993–96 (see Table 4.6).

Table 4.6 Reported issues for NGO advocacy

Period	Combined total number of reported advocacy issues over period	International/Northern NGOs			Oxfam International affiliates		
	A	B	C	D	E	F	G
1981–84	24	17	15	(3) Home country/ Northern aid policies	12	5	(3) Indo-China/ US-led embargo
1985–88	35	20	18	(3) Home country/ Northern aid policies	23	19	(4) Southern Africa/ apartheid
1989–92	45	24	18	(5) Home country/ Northern aid policies	29	21	(4) Home country/ Northern aid policies
1993–96	81	50	41	(6) Debt (4) Landmines (4) Trade	45	33	(6) Debt (6) Home country/ Northern aid policies (4) Trade (5) Landmines (6) African Great Lakes

Notes:

A All of the principal advocacy issues reported by the responding NGOs, recorded at Section 4 of Appendices C1 and C2 in Anderson (2003).
B Total number of reported advocacy issues.
C Number of issues on which only one respondent reported advocacy.
D Issues and numbers of NGOs on which most respondents reported advocacy.
E Total number of reported advocacy issues.
F Number of issues on which only one Oxfam reported advocacy.
G Issues and numbers of Oxfams on which most Oxfams reported advocacy.

Advocacy evaluation

Despite integration of evaluation into the routines of their projects and programmes, NGOs are subject to criticism about shortcomings in evaluation of their work (Tendler 1982; Edwards and Hulme 1992, 1995; Smillie 1995; Sogge 1996b). The need for close and rigorous analysis during the implementation of projects and for more systematic appraisal, monitoring and evaluation are related to a requirement for NGOs 'to build credible alternatives ... to conventional economic thinking' (Edwards and Hulme 1992: 211, 213). A recurrent theme in the literature is the need for more thorough, objective evaluation of NGO effectiveness and the need for the publication of evaluation results as a critical component of NGO transparency (Clark 1991; Edwards and Hulme 1995; Saxby 1996). Similarly, greater attention to evaluation has been stressed as a prerequisite for NGOs to more effectively communicate their advocacy achievements and to win greater private and official donor support for the allocation of resources to advocacy (Edwards and Hulme 1992; Smillie 1995). Therefore, NGOs were asked whether they consistently evaluated their advocacy (or at least claimed to), the bases used for evaluation, and to which stakeholders results were made available.

Less than 50 per cent of the NGOs formally evaluated their advocacy and, of these, only four always did so (Table 4.7). This validates the criticisms of the failure of NGOs to adequately evaluate their advocacy. That only 'occasional' or 'rare' advocacy evaluation was conducted by 13 of 23 respondents (some of which did not have a policy of formal evaluation of their advocacy) is consistent with the widely held view that NGOs are inattentive to evaluating their advocacy and so cannot fully understand its impacts and contribute that understanding to improving future advocacy efforts. Similarly, there is little external release of NGO advocacy evaluations to their stakeholders (Table 4.7), a factor which Clark (1991), Edwards and Hulme (1995) and Saxby (1996) argue should be an essential feature of NGO advocacy and, in Clark's view, to the advantage of NGOs (1991: 73), linking them more closely to broader civil society.

Low levels of evaluation seem to be due to a combination of NGOs having limited advocacy budgets (reflected in survey responses) and the difficulty of evaluating advocacy (Roche 1999; Hudson 2000b; Kelly 2002). Stressing the need for NGOs to demonstrate the effectiveness of this work, while recognising the difficulty in evaluating it, Roche argued:

> A large proportion of advocacy work is long-term; it may lack dramatic moments when it is possible to say that a significant change has occurred. Policy change is often incremental and slow, and implementation lags significantly behind legislative change. Although there may be exceptions to this, particularly in single-issue campaigns, the relationship between these 'victories' and long-term policy change is complex and difficult to untangle. In addition, policy- and decision-making processes are subject to a large number of influences. The necessary change in policy or practice

Table 4.7 Advocacy evaluation (numbers of respondents)

	International/ Northern NGOs	*Oxfam International affiliates*	*Total*
Total respondents	12	11	23
Advocacy formally evaluated	6	5	11
Basis of advocacy evaluation:			
Project or issue	6	5	11
Time	5	4	9
Other basis	2	1	3
Advocacy evaluation consistency:[1]			
Always	2	2	4
Occasionally	6	3	9
Rarely	2	2	4
Advocacy evaluation external release:			
Oxfam International	N/A	2	2
Other Oxfams	N/A	1	1
Funding agencies	4	2	6
Private donors	3	1	4
Southern partners	1	0	1
Researchers	1	1	2
Media	1	0	1

Note:
1 The apparent discrepancy between the number of respondents which advised they do formally evaluate advocacy and the evaluation consistency numbers arises because certain respondents advised they did not formally evaluate advocacy but conducted occasional or rare evaluations.

may happen in a place far removed from where impact is sought; policy and practice changes may need to occur at the local, national, and international level. Combining and cross-checking assessments from multiple levels can be 'like trying to force pieces from different puzzles into one frame' (Oxfam GB: 1998).

(Roche 1999: 193–4)

More current advocacy trends

NGOs which provided data in 1998 were requested to respond in 2005 to a set of generalised questions about current trends in NGO advocacy. These sought insights into the extent to which advocacy had increased in strategic priority and any commensurate changes in financial and human resources dedicated to it; whether NGOs discerned any official funding influences on their advocacy subject matter or policy positions; and the extent to which evaluation of NGOs' advocacy may have increased or decreased since 1996. Agencies were also invited to comment on their advocacy subject matter and the factors they believe limited its effectiveness or which contributed to successful outcomes, and to provide examples of what they considered to have been successful advocacy. Additionally, the Oxfams were requested to indicate the degree to which their collaborative advocacy with Oxfam International and their fellow affiliates had grown or declined in the same period of time.

Perhaps not surprisingly, all 14 responding agencies are now more highly committed to advocacy programmes than prior to 1996. This increased commitment is reflected in significantly increased financial resources being dedicated to policy research, lobbying and public campaigning. While a few of the estimated increases were more modest, only three agencies indicated an increase of less than 100 per cent in resource allocation to advocacy. More than half claimed three- to sixfold increases (compared to 1996) in financial and human resources allocated to advocacy work. Dependence on official funding was 'not a factor' in influencing their advocacy subject matter or their policy position. The two NGOs which offered further comment were both from a G7 country in which NGO incomes are typically heavily reliant on government funding. In each of these cases, their advocacy positions were 'nuanced' or their 'tactics' were influenced by their national government's priorities. One NGO noted: 'Our relationship with the government allows us to be constructively critical when needed, always affirming the positive contribution each of us brings to the work of poverty reduction/global justice.'

NGOs continue to be equivocal and hesitant about evaluating their advocacy. The earlier survey indicated that less than half of the responding NGOs formally evaluated their advocacy; little has changed over the past decade. In 2005, of the 14 responses, half indicated that their advocacy evaluation had increased (in some cases from an acknowledged very low base), while the others indicated that their evaluation remained at about the same level. Since the 14 respondents were broadly representative of the 1998 survey responses (i.e. three indicated they always evaluated their advocacy, four did so occasionally and seven did not), failure to evaluate their advocacy remains a major weakness of NGOs, especially since they are increasingly engaging in it.

Despite failure to consistently evaluate their advocacy, a number of NGOs did not hesitate to list examples of what they considered successful advocacy. This was particularly the case for the Oxfams, which claimed some credit for the successes of Oxfam International-coordinated advocacy. These included the establishment of a direct linkage between World Bank and IMF heavily indebted poor countries' debt relief to poverty reduction programmes in beneficiary countries. Oxfams also cited access to lower-cost generic HIV/AIDS medication and trade policy concessions among their advocacy successes. A number also reported successful advocacy outcomes in relation to their separate initiatives in matters such as fair trade products sold in their own markets, influencing their own governments' aid programmes, contributing to oil and gas revenue-sharing negotiations to the benefit of East Timor, forcing mining companies to change policy and practices and influencing improved rights for low-paid workers in Hong Kong and China. Similarly, the non-Oxfams reported successful outcomes in relation to both global and more localised campaigns and initiatives. As well as identifying their perceived contribution to a number of the successful advocacy initiatives also referred to by the Oxfams, they mentioned influencing their governments' positions in relation to EU economic partnership agreements with poorer countries, ensuring that the Africa Commission report was 'progressive and enlightened', influencing the Canadian government's response to the Boxing Day tsunami, lobbying to maintain strong poverty reduction language in the EU contribution, and the inclusion of key measures in the Australian Senate inquiry into child trafficking.

As many as 30 factors were claimed to have contributed to successful advocacy outcomes. Principal among these were: high-quality and well-grounded research supported by constructive propositions for achievable change (mentioned by several Oxfams); the ability to mount popular campaigns which mobilise large numbers of supporters (which may include activist networks) in combination with good working relationships with key policy decision makers which facilitate lobbying; and working in alliances or coalitions, especially as part of a cohesive and coordinated network with a strong brand. However, only one NGO specially referred to alliances with Southern NGOs which, in the 1998 survey, most agencies indicated they maintained. Media coverage was also mentioned by several respondents, although, surprisingly given their wide use, only one respondent referred to the use of celebrities.

NGOs were also asked to note the factors which limit the effectiveness of their advocacy. The most common limitation was the lack of resources and, therefore, the capacity to develop and implement advocacy programmes. Variously described, this was mentioned by half of the responding agencies. Several NGOs (none of which were Oxfams) acknowledged the lack of evaluation of their advocacy and their inability to attribute outcomes to it as a limiting factor. Several others identified insufficiently developed linkages between their field-based programmes experience and their policy analyses as a constraint on their ability to link causes and effects of poverty. Risk factors were also mentioned by some; these included operating in political environments hostile to NGOs or positions they take (particularly in the US in relation to Palestinians) and the security of field staff working in dangerous situations.

Survey conclusions

The 1998 survey and the 2005 update provide useful insights into Northern NGOs and their use of advocacy as a strategy which, through its contribution to the achievement of the structural and policy macro-reforms necessary to redress the fundamental causes of poverty, is widely regarded as being essential to their poverty eradication goal. The numbers of NGOs which recognise advocacy as a strategy to be employed in pursuit of their objectives, the increasing resources being allocated to it, and the specialised and more senior staff being employed in advocacy suggest that NGOs are heeding the calls for increased priority to be given to advocacy. At least superficially, NGOs are increasingly addressing two of the four NGO advocacy strategic weaknesses: the absence of a clear, coherent strategy and the allocation of resources necessary to effectively implement that strategy; and the failure to build the alliances needed to broaden and strengthen their advocacy voices (Edwards 1993).[5] For the majority of the NGOs, through a combination of the allocation of financial and human resources, recognition of advocacy as a strategy and various alliances, advocacy has been integrated into the fabric of their organisations in pursuit of their poverty reduction and humanitarian relief mission.

The fourth strategic weakness, the 'emasculation' of advocacy for fear of reductions in official funding, on which many are so dependent (Edwards 1993: 168), was not substantiated and was specifically rejected by the NGOs. The lack of correlation between official funding and advocacy expenditures and, indeed, the contradictions reported above suggest organisational culture and its priorities rather than reliance on official funding determine the emphasis placed on advocacy and the resources allocated to it. While no correlation between official funding and advocacy expenditures was found, it was beyond the scope of the survey to examine the nature of the advocacy and the extent to which its messages may be influenced by dependence on official donors. Thus, it is possible that content, rather than the decision to engage in advocacy, may be influenced by official donor funding dependency (Minear 1987: 207). Another possibility is that content may, of course, be influenced by the particular agency's constituency; for example, several of the Christian agencies indicated a theological basis for their advocacy. In 2005, one Christian agency commented: 'Our faith-based approach is a source of strength in our advocacy work. A Christian approach to rights-based development, articulated and used as an ongoing frame for our work, gives coherence to our advocacy.'

A further major NGO advocacy weakness is the failure of NGOs to demonstrate to themselves and their stakeholders, through evaluation, the effectiveness of their advocacy as justification for the financial and human resources dedicated to it (Clark 1991; Edwards and Hulme 1995; Saxby 1996). Evaluation, documentation and publication of advocacy experience will, in addition to assisting in demonstrating the effectiveness of NGOs' advocacy and their accountability, 'facilitate scaling up by others' (Edwards and Hulme 1992, 1995: 224; Archer 1994: 232). Without the foundation provided by consistent, thorough evaluation of their advocacy, NGOs will be unable to assess its effectiveness and address and, if necessary, redress the criticisms made of their advocacy. Furthermore, NGOs will be unable to realise the macro-reforms acknowledged to be essential to advancing their cause of impacting substantially on worldwide poverty and related injustice (see Hudson 2001a, 2001b). Until NGOs themselves have become sufficiently confident in the effectiveness of their advocacy to communicate and demonstrate its achievements, it will surely remain an under-utilised and resourced component of NGO strategy, notwithstanding its potential contribution to their mission. If consistent, thorough evaluation of their advocacy is a prerequisite for such a level of informed confidence, survey responses suggest that much greater priority be given to advocacy evaluation by NGOs.

The following section of this chapter presents a case study. It explores the ways in which a group of previously loosely connected international NGOs formalised their relationship to form Oxfam International, one of the principal purposes being to 'scale up' their separate and collective advocacy capacity.

Oxfam International and its advocacy purpose

Stichting Oxfam International (Oxfam International, OI or Oxfam) was formally established as a charitable foundation in the Netherlands on 2 January 1996. Oxfam International was formed with nine founder affiliates (or Oxfams). These development and relief NGOs were established as national organisations in each of the countries in which they were formed, with the exception of Canada where there are two, Oxfam Canada and Oxfam Quebec. Oxfam Hong Kong, which has China as its principal geographical programme focus, is head-quartered in Hong Kong as a special administrative region of China. All are based in Northern, developed economies, and only one, Oxfam Hong Kong, is not in an OECD member country. These autonomous affiliates share a common vision and philosophy and similar working practices. Prior to the formation of Oxfam International they had worked in a loose association 'challenging inequality and injustice by working around the world to overcome suffering caused by poverty, conflict and natural disasters' (OI 1999a: 1). The globalisation of the world economic and political systems influenced the Oxfams to strengthen their 'international collaboration . . . to make [a] real impact on poverty and injustice world wide' (OI 1999a: 1).

As one of the principal objectives of Oxfam International, Article 2b of its Constitution provides for OI to 'to research the causes and effects of poverty, distress and suffering and to educate the general public and decision makers as to same'. Article 5 of the Constitution, in detailing the means Oxfam International is to use to achieve its objectives, provides that it shall 'facilitate international advocacy, research and policy development; and provide public education' (OI 1996).

Oxfam International established its Washington Advocacy Office (WAO) in 1995 to 'co-ordinate the development of joint strategies and policy positions for the Oxfams' (OI 1999b). Washington, DC was selected as the location of the OI advocacy office because of the perceived strategic necessity to have direct and effective access to the Washington-based multilateral agencies (MLAs), the World Bank, the International Monetary Fund (IMF), and the United Nations (UN) based in New York.

Oxfam International is an international group of independent non-governmental organizations dedicated to fighting poverty and injustice around the world. The Oxfams work together internationally to achieve greater impact by their collective efforts.

(Extract from Oxfam International's Mission Statement)

Figure 4.1 Member organisations of Oxfam

Source: OI 2000: 2.

Establishment of the Washington Advocacy Office

Establishment of the WAO had its genesis in early 1993 discussions between the then international liaison officer for the nascent OI and representatives of several national Oxfams. These discussions were referred to later in 'International advocacy office operational and management paper' (the Spring

1994 paper) as being 'around ways the Oxfams could have closer collaboration that would result in more consistent and unified advocacy efforts and that would have clearer, more tangible impact on global issues and international polices that affect partners' (OI 1994a: 2).

As a result of these early 1993 discussions, a draft paper was developed by Oxfam America for an international advocacy workshop held in June 1993 and attended by representatives of the then loosely collaborating Oxfams. That meeting agreed, in principle, to the formation of a joint international advocacy office and for participants to take the proposal back to their respective Oxfams for further discussion. A resulting paper (the September 1993 paper or proposal) set out the rationale for establishing an international advocacy office, reviewed the role of international organisations in addressing the needs and concerns of the 'South' and the role that the 'family' of Oxfams might have in influencing relevant policies, and outlined the objectives and possible locations, staffing structure, accountability arrangements and costs of the office (OI 1993a). Fundamental to the proposal was the globalisation of the political economy, largely shaped by the developed economies of the North and reinforced by communication technologies. In outlining the rationale for the establishment of an international advocacy office, the paper stated:

> . . . agencies seeking meaningful policy analysis and formulation require direct and urgent access to considerable and diverse information on issues, since decisions and subsequent actions are determined in large part by the very speed of communication and the interlocking of concerns and interests of major global actors.
>
> The reshaping of a 'new world order' to the next century must urgently address and develop clear positions on such questions as widespread and increasing poverty; continuing inequitable resource allocation and 'uneven development' among peoples; increasing proclivity among many countries to major disasters coupled with apparently reduced capacity to address them from their own resources; etc. There are essential global concerns calling for global analysis and action.
>
> (OI 1993a: 3)

And, in relation to the role of the Oxfams in addressing these issues:

> The Oxfam agencies, if they are to fulfil their mission and exercise influence on the change process, must be present in these instances where they can both represent their Northern constituencies and their partnerships with peer agencies of the South, to provide a forum for their effective participation in international debates. . . .
>
> The complex and dynamic nature of the development debate at this critical juncture of history requires that agencies engage in promoting fundamental change, [and] be even more strategically located in relation to the loci of international policy analysis, formulation and action.
>
> (OI 1993a: 5, 6)

The objectives of the office, which 'should be seen as representing the totality of the Oxfam family . . . regardless of their financial contributions', were at that time directed towards being a means of closer collaboration and a source of information relevant to 'policies that impact international organisations and efforts' (OI 1993a: 7).

The location of the office received considerable attention, with suggestions that it be located in either New York, the location of the United Nations secretariat headquarters and key agencies, or Washington, DC, as the headquarters of international financial institutions and the base for a number of coalitions and research groups concerned with international development (OI 1993a). Various Oxfams generally favoured Washington, DC as providing more direct access to the World Bank, the IMF and the United States government while being relatively close to New York.

Although the proposal for the establishment of an international advocacy office was sponsored by Oxfam America, support from Oxfam Great Britain (then Oxfam United Kingdom and Ireland) was a key factor. As the largest Oxfam, it was already active in international advocacy. Based on its 1993 analysis of world poverty, Oxfam United Kingdom and Ireland resolved to materially expand the resources allocated to advocacy and to 'develop from being a purely UKI organisation to one which is both more European and more truly international . . . in association with our sister Oxfams' (Oxfam United Kingdom and Ireland 1993: 7).

The September 1993 meeting of executive directors (chief executives) of the Oxfams agreed to recommend the establishment of 'International Oxfam Advocacy Office in Washington but taking New York and the UN into account' and, further, that 'the issue of advocacy and popular mobilisation' should be included on the agenda for the international conference of Oxfams to be held in May 1994 (OI 1993b: 17, 18). The Spring 1994 paper was circulated to all Oxfams for discussion and 'endorsement at relevant levels' before consideration by the executive directors and chairs at their international conference of Oxfams held in May 1994 (OI 1994a: 2). Although concerned primarily with operational and management matters, this paper made several new points in relation to the international advocacy strategy: the International Advocacy Office (as it was then known) would be a vehicle for maximising the 'effect of the collective positions of the International Oxfam Family on agreed upon global issues policies and strategies directed at targeted international institutions'; the office as an 'outgrowth' of the national advocacy efforts of each Oxfam was to be supported by Oxfams as an integral part of their own advocacy; and there was strong consensus 'that if the Oxfams are to fulfil their missions and exercise influence on the change process, we must be viewed as one strong voice, present and prepared to represent all our constituencies both North and South'. The importance of the advocacy office as 'an important building block for the Oxfam Family to work towards a larger good, the creation of Oxfam International', was also recognised (OI 1994a: 3).

The Spring 1994 paper also reflected a significant evolution of the essentially information-gathering, representation and preliminary analysis objectives and functions of the office, compared to those envisaged earlier. The aims of the office were intended to be more directly advocacy-oriented, to change policies on 'targeted issues of concern to the Oxfam Family and its partners' and included 'building on and advancing the advocacy experiences, efforts and objectives of the Oxfams' (OI 1994a: 3). Functions too were much more directly advocacy related than previously contemplated, including lobbying, coordinating the implementation of agreed strategies, facilitating [Southern] partner and Oxfam visits, and working with other organisations and networks. The office's role in developing advocacy strategies was emphasised, involving facilitation of policy and strategy development and strategic thinking on issues and political process in relation to the World Bank, the IMF and the United Nations (OI 1994a: 3, 4). New operational plans were also set out, including advocacy issue selection criteria, an annual workplan development process (including proposals for specific projects for the first year of operation), proposals for the conduct of an annual advocacy workshop, interim (until establishment of Oxfam International) management arrangements and staffing proposals agreed by the steering committee. The advocacy issue selection criteria emphasised that advocacy would be informed by field experience, linked to the World Bank, the IMF or the UN, and provide the opportunity for Southern partner influence. Networking with other NGOs and their networks and coalitions was also proposed (OI 1994a: 5). These proposals were approved by the executive directors and chairs at the 1994 Oxfam International conference, and the Oxfam International Advocacy Office commenced operations in Washington, DC in February 1995.

Washington Advocacy Office governance, management and staffing

Governance[6] of the WAO is exercised by the Board of OI, and operationalised primarily through the Council of Executive Directors composed of the chief executives of each of the OI affiliates. Line management of OI is through the executive director of Oxfam International (OIED), who is based in the Oxford UK Secretariat office. Management of the operational programme of the WAO was, until 2003, coordinated through and agreed with the OI Advocacy Coordinating Committee (ACC) chaired by the OIED. Until OI's operations were restructured, the ACC comprised a senior staff member of each of the affiliates with responsibility for advocacy within their respective affiliates. The organisational structure of Oxfam International, including the WAO, is presented in Figure 4.2.[7]

Over time, the WAO workplan evolved; after initially being considered by the OI Board, the advocacy strategy was incorporated into the OI Strategic Plan 1999–2000 and coordination delegated to the ACC.

As one of four OI coordinating committees which operated until early 2003, the ACC was responsible for planning and coordinating OI's advocacy programme and advising the Council of Executive Directors (OI 1999c). The

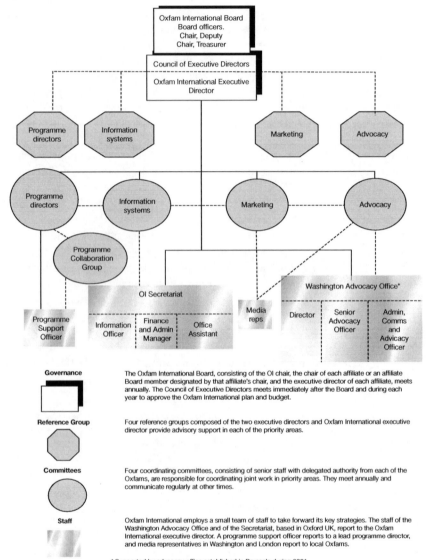

Figure 4.2 Oxfam organisation structure to 2002

Source: OI 1999b.

ACC was therefore the primary operational link between OI affiliates and the WAO and between their own advocacy programmes and those of OI. ACC members representing the affiliates are responsible to affiliate executive directors, all of whom are members of the Council of Executive Directors, providing a means of effective coordination of strategic and operational management.

One function of the ACC was approval of public statements issued in the name of Oxfam International relating to the development and humanitarian emergency programmes of affiliates and the WAO. Policy requires that advocacy statements issued in the name of OI should have the unanimous support of all affiliates obtained through their ACC representatives. If a unanimous decision cannot be obtained, a vote by executive directors is held and a decision can be blocked by two votes. Thus, OI seeks affiliate unanimity, but also provides a mechanism for issuing public advocacy statements where that unanimity is not possible (OI 1997). OI aims at a comprehensive, inclusive, global vision.

OI's Washington-based advocacy director is responsible for the management of OI's advocacy programme. During 2001 and 2002, advocacy offices reporting to the advocacy director were established in Brussels, Geneva and New York. The Brussels office is primarily concerned with European Union-oriented advocacy, Geneva is focused particularly on the World Trade Organization, and New York is focused on the United Nations. OI's international advocacy thus became more focused and increasingly strategic in orientation.

As part of a major realignment of OI's operational structure, in early 2003 certain ACC responsibilities were transferred to a Global Coordination Team (GCT) which monitors progress against OI's Strategic Plan, recommends global priorities, and generally coordinates OI and collaborative programmes between affiliates (OI 2002a). A 'virtual' ACC continues to approve advocacy statements issued in OI's name (OI 2002a). Thus, formal processes facilitate careful monitoring of OI's global activities.

Washington Advocacy Office funding and spending

In 1995 when affiliates contributed to establishment of the WAO, its operating costs were funded by contributions from these affiliates in amounts unrelated to their usual contribution to the costs of operating OI. Each of the affiliates decided what amount they would contribute to those first-year costs without reference to the established contribution formula. One affiliate, Oxfam Hong Kong, underwrote the project by offering to contribute any shortfall of contributions compared to budgeted first-year costs up to $US200,000 (OI 1994a, 1994b). Since the WAO's first year of operations, its costs have been funded as part of the OI budget covering both the OI Secretariat and WAO operating costs. Affiliates contribute to these costs under an agreed formula which has regard to affiliates' income levels. Additional funds to support OI's WAO-led advocacy were made available through a combination of those affiliate contributions, an external fundraising programme during 1998 and 1999, and a development investment fund to which contributions by Oxfam Hong Kong and a Board member supported the strategic development of OI in its formative years.

Until 1999, spending on the WAO increased at the rate of approximately 12 per cent per annum from a budgeted maximum US$172,400 operating cost for its first full year of operation (OI 1994a, 1999d). The proportion of expenditure on employment and related costs has scarcely varied over the years, constituting 68.4 per cent of the first-year costs and 66.9 per cent of budgeted 1999 expenditures (OI 1994a, 1999d). By 2002, the WAO's annual budget had increased to $US393,897 (OI 2002b), and budgeted expenditure for 2003 rose to $US412,633 (OI 2003). Additionally, various OI advocacy activities, including costs of the Geneva and Brussels offices and a Washington-based media staff member, were funded by affiliates. Further expenditure was also committed to OI's trade campaign which commenced in 2002 and was funded by affiliates. By 2002, OI advocacy expenditures had risen more than fourfold. When added to affiliates' own spending on OI-coordinated advocacy, the financial resources allocated to the OI-led collaborative advocacy programme are indicative of the importance attached by affiliates to advocacy and of advocacy's centrality to the overall OI strategy.

'Scaling up' OI advocacy

Establishing the WAO in 1995 was a major strategic step taken by the nine Oxfams, alongside the formal establishment of OI in that year, with collaborative advocacy being central to its overall strategy. This increased advocacy priority is reflected in OI's Strategic Plan 2001–2004 (later extended to 2006). The foreword to the Strategic Plan states:

> As Oxfam International, we cannot achieve our ambitious objectives on our own, however much we grow. Working in alliances with other networks and organisations, promoting a broad movement of global citizens for social justice will be a fundamental aim. Strengthening Oxfam's own capacity to make the strongest possible contribution to these alliances is crucial. It will be part of all dimensions of our activities in program, humanitarian response, advocacy and marketing.
>
> (OI 2001a)

The Strategic Plan commits OI to 'significantly improve the quality, efficiency and coherence' of its work (OI 2001b; see also OI 2001c). This includes collaborating more fully with allies as part of a global campaigning force; reflecting Oxfam's core values in its communications; achieving optimum effectiveness through alignment of the programmes and resources of affiliates; and increasing Oxfam's advocacy effectiveness.

Both the 1998 and the 2005 survey revealed increased advocacy collaboration between OI affiliates and, through OI's advocacy offices, a high level of cohesion in Oxfam's global, regional and national advocacy. Affiliates of OI, the WAO and its more recently established European offices are central to Oxfam's advocacy strategy. This cohesion is reflected in the efforts of OI and its national affiliates to influence the World Bank's Heavily Indebted Poor Countries Initiative, which, to illustrate the 'scaling up' and globalisation of advocacy, are reviewed in the following chapter.

5   Oxfam, the World Bank and heavily indebted poor countries

Ian Anderson

Oxfam International's statement of its mission in pursuit of global poverty reduction is explicit in its commitment to addressing the structural causes of poverty. This places OI's advocacy at the heart of its strategy, central to which is its Washington Advocacy Office (WAO), which coordinates Oxfam's advocacy globally. OI's first advocacy office was located in Washington, DC in 1995 because of the priority then attached to influencing the World Bank and IMF. Globally the World Bank is the largest provider of development finance, so influencing World Bank policies and practice was, at that time, a key element in OI's advocacy strategy.

Although the World Bank and Oxfam share the same fundamental goal of poverty reduction, relationships between the two organisations are complex, reflecting their differing constituencies and priorities. As an international financial institution, the World Bank formulates and determines policy through a complex set of political and economic interactions between its shareholders, shareholder-appointed governors and executive directors, and management and operational staff. Oxfam International is grounded in the civil societies of each of its 13 affiliates.[1] Despite substantial similarities in their respective mission statements, the World Bank and Oxfam do not always share the same values, perspectives and objectives.

Because of the centrality of the World Bank to Oxfam's advocacy strategy and the sustained intensity of World Bank-oriented advocacy in the WAO's earlier years, it was singled out for case study research as a basis for evaluating the policy outcomes of Northern NGO advocacy. OI's World Bank-oriented debt relief advocacy in relation to the Heavily Indebted Poor Countries Initiative (HIPC) illustrates one means by which Oxfam (itself or in collaboration or alliance with other NGOs) sought to influence World Bank policy and, thereby, practice.[2]

Two methods were used to gather data: the first, a document-based review, provided evidence of the respective positions of the World Bank and Oxfam, and the second involved a series of semi-structured interviews with World Bank, Oxfam and other NGO management and advocacy staff, political figures who were engaged in the HIPC Initiative, and journalists who had covered it. World

Bank personnel interviewed represented shareholder-appointed executive directors, senior management (including the president as chief executive), and specialist HIPC staff who had engaged with Oxfam. Interviews were mainly conducted in Washington in late 2002. The review traced the evolution of relevant World Bank policy, Oxfam's advocacy objectives, and its analysis of World Bank policy from commencement through to 2001. The interviews complemented this by providing informed insights into complex policy formulation processes not apparent from publicly available and unpublished policy documents, position papers, official communiqués and media releases. This chapter thus reviews Oxfam's efforts to influence the World Bank's HIPC policy and policy outcomes, particularly as it relates to strengthening the link between debt relief and poverty reduction programmes in the HIPC beneficiary countries. This illustrates the capacity of one NGO to influence the policies of an international financial institution and suggests other NGOs may have similar ability – vitally important to NGOs achieving their poverty reduction goals.

Oxfam and the HIPC initiative: a historical review

In 1996, Oxfam's debt relief advocacy objective was to encourage a 'Comprehensive, coordinated, sustainable, solution to [the] debt problem' (ACC 1996). Over time, this objective was refined to winning support from the World Bank and key creditors for relief to countries which demonstrated a willingness to commit resources to concrete poverty reduction programmes (ACC 1998). Oxfam continued to emphasise the link between debt relief and poverty reduction by pressing for 'radical changes to HIPC by establishing [a] human development window to give deeper and quicker debt reduction to countries committed to poverty reduction' (ACC 1999). Oxfam's advocacy strategies involved a mutually reinforcing combination of the issue of numerous position papers and briefings, lobbying, and public campaigning, supported by media coverage. This activity was underpinned by research which linked debt and poverty in the HIPC countries, in relation to which several country-specific plans were developed. In its lobbying, Oxfam developed plans for working with Southern organisations and for coordinating its Washington-based efforts with those of OI affiliates in key G7 creditor countries. Advocacy strategies also included working collaboratively with other NGOs and NGO coalitions including the Jubilee 2000 network, EURODAD and Social Watch (ACC 1998). As well as efforts directed towards key creditor countries (especially G7 governments), the World Bank was a key focus of Oxfam's debt-related advocacy as Oxfam and its partner NGOs pressed it to take a more proactive and political role to 'encourage debtor governments to make commitments on how they will use debt relief' (ACC 1998: 5).

Oxfam's emphasis on the link between debt relief granted to less economically developed countries, under the joint World Bank and IMF Initiative for Heavily Indebted Poor Countries (HIPC Initiative), and poverty reduction-oriented social spending was a distinctive component of its World Bank-focused advocacy. Oxfam used the term 'Human Development Window' (HDW) to emphasise that governments which committed to using financial resources created through HIPC relief for poverty reduction programmes should receive 'earlier and deeper debt relief' (OI 1998b: 2).[3]

The HIPC Initiative had its origins in the Halifax G7 Summit, of June 1995, which called for a comprehensive approach to resolving the unsustainable debt problem of the world's poorest and most indebted countries. The initiative was endorsed by the World Bank and IMF and approved for implementation in 1996 (World Bank Group 1998a).[4] Oxfam's Human Development Window proposals grew out of its advocacy in response to the HIPC Initiative.

The opportunities for linking debt relief with poverty reduction were first raised by Oxfam in a September 1996 position paper, and in April 1997 Oxfam challenged the World Bank president to implement HIPC in order to, in the opening words of the World Bank's *Strategic Compact*, 'achieve greater effectiveness in the World Bank's basic mission – poverty reduction' (OI 1997c: 4). Under its proposals for 'the adoption of a new debt-for-poverty reduction contract in the HIPC initiative', Oxfam argued: 'Poverty reduction incentives should be integrated into the HIPC framework. Countries willing to engage in a dialogue aimed at converting debt relief into poverty reduction initiatives should be rewarded with an accelerated time-frame for debt relief' (OI 1997c: 7). Oxfam also noted that the existing HIPC framework failed to 'consider human welfare as a factor in determining debt sustainability' and, in assessing the time-frame and scale of debt relief, had not considered 'the potential for achieving social development gains' as a determining factor (OI 1997c: 27). Consequently, under its 'debt-for-poverty reduction contract' proposals to be 'negotiated between debtors and creditors with the active participation of civil society, donors and NGOs', Oxfam pressed for poverty reduction as one of the 'central purposes of the HIPC initiative'. Oxfam argued this would 'provide incentives for converting debt relief into human welfare gains and introduce positive conditionalities aimed at reinforcing those incentives' (OI 1997c: 27).

A year later, Oxfam referred to a 'new poverty reduction window', proposing that governments which committed to using resources created through HIPC relief for poverty reduction programmes should be provided with 'earlier and deeper debt relief' (OI 1998a: 2). Oxfam's September 1998 position paper contributed to the HIPC review, then being undertaken jointly by the World Bank and the IMF, and proposed a system of incentives in the form of earlier and deeper debt relief for countries willing to make commitments to poverty reduction (OI 1998b: 1).

On 22 September 1998, staff of the World Bank and IMF released a paper for the Development Committee (a forum of the World Bank and the International Monetary Fund[5]) which stated: 'the HIPC framework has always emphasized the need to link debt reduction with effective long-term policies for economic and social development and poverty alleviation' (World Bank Group 1998a: 15). It acknowledged calls for 'debt-relief efforts to be explicitly integrated into a broader strategy to combat poverty' (World Bank Group 1998a: 48, 49).

However, it largely dismissed them as being unnecessary in view of the HIPC emphasis on economic and social development and the fact that 'social development criteria are developed jointly with country authorities and are explicitly incorporated into HIPC conditionality' (World Bank Group 1998a: 50). Nearly one year later the IMF and the IDA[6] published an HIPC review paper acknowledging extensive consultation with civil society, which 'focused on the link between debt relief and poverty reduction' (World Bank Group 1999a: 2).

Drawing on published and unpublished documents, this review of Oxfam's World Bank-oriented HIPC advocacy traces the evolution of Oxfam's propositions and the World Bank's responses. It also outlines the development of World Bank policy related to linkages between HIPC relief and poverty reduction-related social spending by beneficiary countries.

Linking HIPC relief and social expenditure: World Bank policy overview

Although the World Bank recognised an indirect link between HIPC relief and poverty alleviation in the context of 'long term policies for economic and social development', it rejected a direct relationship between debt relief and social expenditures (World Bank Group 1998a: 15, 1998b: 1). Its reasons included: that countries' social expenditure needs differ from the cash flow benefits from HIPC relief, especially where debt service obligations had not been paid; social expenditure programmes' absorption capacity constraints; and that most HIPCs already receive substantial development assistance for which debt relief should not be substituted. Rather, the World Bank (and IMF) argued that the link between HIPC relief and social development 'should be viewed in the broader perspective of the overall poverty alleviation efforts supported by creditors and donors through various instruments' (World Bank Group 1998a: 15). Despite downplaying any more direct link between debt relief and poverty reduction and social investment programmes as proposed under Oxfam's Human Development Window, throughout 1999 substantial movement occurred in this aspect of World Bank HIPC policy.

In February 1999, the World Bank and IMF launched their 1999 HIPC review directed towards making 'HIPC as effective as it can possibly be' (IMF 1999: 1). Recognising the value of consultations with civil society throughout the HIPC process, the 1999 HIPC review sought further input from civil society organisations. The first phase invited views on technical issues of debt sustainability, time-frames and links to macro-economic and structural policy reforms. These were to be reported to the Boards of the World Bank and IMF leading up to the meetings of the Interim and Development Committees in late April 1999. The second phase, directed towards reporting to both institutions' Boards prior to their September 1999 annual meetings, sought views 'on the relationship between debt relief, social policies and poverty reduction' (IMF 1999: 1). The poverty reduction questions to be addressed were:

> How can the link between poverty reduction and debt relief be strengthened in the programs supported through the HIPC Initiative? How should they be linked to the achievement of the international development goals set for 2015? How can the debt relief provided be most effectively used to foster social development particularly in the health and education fields?
>
> (IMF 1999: 3)

A related paper acknowledged 'the unprecedented degree of helpful participation and coordination between all actors involved – debtor governments, creditors, and donors – as well as the intense interest shown by NGOs, churches, and other groups in the Initiative' (World Bank Group 1999b: 1). Shortly thereafter, the World Bank's *Development News* reported the Canadian prime minister's proposals for more generous debt relief, acknowledging they were being made as the HIPC review was taking place and 'as prominent NGOs like Jubilee 2000 and Oxfam International are generating unprecedented interest in this important subject' (World Bank Group 1999c: 1).

The Phase I consultation was fully reported in the World Bank's HIPC review paper of April 1999, which included extended summaries of proposals from a number of the contributing government, religious and civil society organisations. Oxfam's proposals, presented in Box 5.1, were also further acknowledged by the World Bank:

> Oxfam has made a concrete proposal for establishing a Human Development Window that would provide additional benefit to countries which make a credible commitment to target funds released from debt service to meeting these social sector needs. This topic will be the main focus of Phase II, in which Bank and Fund staff will explore, with external input, ways of strengthening this link and will report back on this process before the annual meetings.
>
> (World Bank Group 1999d: 9)

A few days later, the manager of the World Bank's HIPC unit responsible for leading its part in the review acknowledged the role of Jubilee 2000 and its members, including Oxfam: 'Jubilee 2000, the worldwide movement to cancel the international debt of poor countries by the new millennium, has provoked a necessary and healthy debate and has played a critical role in forcing a discussion in industrialized countries on debt' (World Bank Group 1999e: 6).

The World Bank and IMF's policy shift towards strengthening the link between debt relief and poverty reduction was reflected in a progress report of 21 April, which outlined the first of the principles for changing HIPC: 'Debt relief should be provided in a way that reinforces the wider tools of the international community to promote sustainable development and poverty reduction' (World Bank Group 1999g: 2). Later, in applying these principles, the IMF managing director and World Bank president added that:

We share the widespread interest in ensuring that the resources released through debt relief contribute to strengthening programs to improve social and economic services for the poor. Debt relief can provide both an incentive and a relatively sustained resource to back poverty reduction and social development programs. We believe that there is scope to tighten these linkages within the Initiative. A focus of the second stage of the 1999 HIPC review process now underway is to improve the mechanism to achieve this.

(World Bank Group 1999f: 2)

Box 5.1: **Oxfam's debt relief and poverty reduction: strengthening the linkage**

Oxfam advocates minimum debt levels as follows:

- Debt service ratio should be reduced to 15–20 per cent.
- The net present value (NPV) of debt to exports ratio should be reduced to between 150 and 200 per cent.
- The NPV of debt to government revenues ratio should be reduced to 200 per cent.
- The time-frame needs to be reduced from six to three years.
- The fiscal criteria of government revenues to GDP should be reduced to 15 per cent.
- The debt service paid by the government should be restricted to 10 per cent of government revenue.

In addition, to link debt relief further to development, Oxfam suggests putting incentives in place to reward governments willing to 'enter into genuine poverty reduction partnerships' by providing earlier and deeper debt relief through their Human Development Window, where thresholds can be further lowered to:

- a debt service ratio of 10–15 per cent;
- the NPV of debt to exports ratio between 100 and 150 per cent;
- the NPV of debt to government revenues ratio between 150 and 170 per cent.

The additional assistance could not be provided unless a government is willing to commit 85–100 per cent of savings on debt service to identified poverty reduction initiatives. The Human Development Window could be one of the instruments to address the challenges to fund education in HIPCs, as elaborated in *Education Now! Break the Cycle of Poverty*, published in March 1999 (OI 1999f).

(Source: World Bank Group 1999d: 6)

Concluding remarks by the acting chairman of the World Bank Executive Board highlighted the strengthened link between debt relief and poverty reduction, seeming to endorse Oxfam's proposal to give HIPCs incentives to invest in poverty reduction programmes:

> On the links between debt relief and poverty reduction and social policies, Directors considered that the resources released through debt relief should contribute to improving social and economic services for the poor. Debt relief can provide both an incentive and a resource to support poverty reduction and social development programmes. Directors welcomed the intention to consider ways to tighten the linkages between debt relief within the Initiative and poverty reduction and social policies in the context of the second part of the 1999 HIPC Initiative consultation and review process. They emphasised that such work would need to rely heavily on the expertise of the World Bank.
>
> (World Bank Group 1999f: 8)

In July 1999 the IMF and World Bank's IDA released proposals for modifications to the HIPC Initiative. In general terms, the proposals provided for deeper, earlier and broader (to more countries) debt relief. They also acknowledged the need for debt relief to promote the wider goals of poverty reduction and sustainable development and foreshadowed a further paper containing proposals for an 'enhanced framework for poverty reduction' (World Bank Group 1999g: 23).

Shortly after, the World Bank participated in an HIPC review seminar in Addis Ababa, Ethiopia, hosted by the United Nations Economic Commission for Africa. The seminar was also attended by representatives of 22 HIPCs, a number of G7 and OECD member countries, the IMF and leading NGOs. Foremost among the seminar's outcomes was its endorsement of the need for a strong link between debt relief and poverty reduction. It was acknowledged this link would require a combination of an appropriate macro-economic framework and country-led and broad-based poverty reduction strategies, of which debt relief was an integral part (UNECA 1999).

In August 1999, the IMF and IDA released the HIPC review Phase II paper. This proposed that 'The main elements of the poverty reduction strategy would be published as a tripartite document endorsed by the government, the Bank and the Fund – Poverty Reduction Strategy Paper (PRSP)' (World Bank Group 1999h: 29). This appears to be the first published use of the term 'PRSP'. The PRSP was, in two major respects, conceptually similar to Oxfam's proposals. Firstly, the Phase II paper reported that 'Oxfam has suggested that the government submit a poverty action framework at, or in advance of, the country's decision point under the Initiative and that this should serve as a "contract" between creditors and the debtor government' (World Bank Group 1999h: 9). Secondly, the proposed PRSP specifically provided for 'consultations with civil society and other stakeholders' for which Oxfam (whose proposals were specifically acknowledged along with those from World Vision) and other NGOs had argued. This was seen as a means of ensuring the strategy was appropriate to the country's circumstances and the 'most appropriate means of fostering ownership of poverty reduction strategies and accountability for the use of resources available for their financing' (World Bank Group 1999h: 9–10).

Considered by the Development Committee, these proposals put forward an 'enhanced framework' which contained three key elements: 'A comprehensive understanding of poverty [in the relevant country] and its determinants'; 'choosing actions that have the highest poverty impact'; and 'outcome indicators which are set and monitored using participating processes' (World Bank Group 1999i: 2–3). These elements are 'part of a broader effort to enhance the impact of the Bank's work on poverty reduction to promote sustainable economic growth and to ensure the benefits are reaching the poor' (World Bank Group 1999i: 3). The creation of PRSPs, within the framework proposed, was approved in the Development Committee's communiqué of 27 September 1999 (World Bank Group 1999j).

Oxfam's HIPC advocacy

Oxfam's aim to link poverty reduction with debt relief sought to win support from the World Bank and from key creditors for countries that demonstrated a willingness to commit resources to concrete poverty reduction programmes. Oxfam's 1999–2001 Workplan stated that its objective was to achieve 'radical changes to HIPC by establishing [a] human development window to give deeper and quicker debt reduction to countries committed to poverty reduction' (ACC 1999). An additional objective was to challenge the limitations of the time-frame, sustainability levels, and the linkage with IMF conditions contained in the HIPC framework (ACC 1998).

Oxfam had been actively engaged in the HIPC Initiative since it was first proposed. Initial involvement concentrated on the need for a comprehensive strategy to address the multilateral debt burden of what were then being referred to as 'severely indebted low income countries' (OI 1995: 1). Until April 1997, when Oxfam challenged the World Bank to take seriously its poverty reduction mission in implementing the HIPC Initiative, its analysis concentrated on the eligibility for and scale of relief, debt sustainability levels and the gap between the decision and completion points in relation to HIPC relief.

In its April 1997 challenge to the World Bank and its proposal for a 'debt-for-poverty-reduction contract', Oxfam outlined an agenda for reform which had at its heart human welfare and HIPC's potential for contributing to social development. Oxfam pressed for acceleration of the HIPC time-frame, the lowering and broadening of debt sustainability thresholds, more weight being given to HIPC's fiscal considerations, and abandonment of IMF conditionality. In addition, it argued: 'Poverty reduction should be established as a central objective of the HIPC initiative with human development indicators being given more prominence in determining eligibility' (OI 1997c: 30). This proposed centrality of poverty reduction to the HIPC Initiative was supported by Oxfam's five-phase plan, which involved:

1   joint reviews of existing social policy initiatives by governments, creditors and UNICEF to examine performance against targets for human development indicators;
2   dialogue between governments, donors, NGOs and civil society over priority areas and targets for social welfare achievements;
3   estimates of the additional social investment resources which could be generated through debt reduction to bridge the gap between domestic finance and the budgets required to achieve designated human welfare targets;
4   identifying specific social investments to be financed through the resources generated by debt relief;
5   drawing up a contract setting out specific government commitments for converting debt relief into poverty reduction measures, including the timetable for spending allocations and benchmarks for measuring progress and performance.

(OI 1997c: 27–8)

Contrary to IMF claims, Oxfam argued that the evidence showed that 'debt is a poverty issue – and that debt relief *can* contribute to poverty reduction' (OI 1997c: 28). Oxfam linked the incentive for accelerated relief to HIPCs being 'willing to engage in a dialogue aimed at converting debt relief into poverty reduction initiatives' (OI 1997c: 7).

A year on, the HIPC Initiative was seen by Oxfam as faltering because of a lack of political will on the part of the World Bank, IMF and their major G7 shareholders to fully implement it. Oxfam gave priority to revitalising the initiative, strengthening its recommended conditions for faster and greater debt relief, and proposing that those benefits be given to HIPC governments which 'commit to use savings from debt servicing for poverty reduction initiatives' (OI 1998b: 2).

This strengthened link between debt relief and poverty reduction programmes in HIPC countries was further developed by August 1998. As a contribution to the HIPC Phase II review, Oxfam set out 'a proposal for developing a more poverty-focused approach to debt relief' and introduced proposals for lower debt sustainability thresholds:

•   the HIPC Initiative to be integrated into the evolving global strategy for poverty reduction set out by the Development Assistance Committee of the OECD;
•   the creation of a debt for human development window, under which governments willing to allocate 85–100 per cent of the savings from debt to poverty reduction initiatives would be given improved incentives in the form of earlier and deeper debt relief;

- cooperation between debtor governments, donors and civil society in developing poverty action frameworks through which resources released through debt relief can be channelled;
- the development of a transparent and accountable structure for administering the debt for human development window, modelled on the precedent set by the Ugandan government.

(OI 1998a: 1–2)

Oxfam followed this almost immediately with a short commentary on the World Bank/IMF HIPC review paper of 22 September 1998, styling it as 'a wasted opportunity' and listing five principal areas where HIPC was 'falling short':

1   The debt sustainability ratios were too high;
2   The time-frame for debt reduction was unnecessarily long;
3   There was a lack of flexibility for post-conflict countries;
4   Eligibility for debt reduction was too rigidly linked to an IMF Enhanced Structural Adjustment Facility (ESAF) track record;
5   There was an inadequate connection between debt relief and poverty reduction.

(OI 1998c: 2)

Oxfam again called for the establishment of a 'human development window' to provide a means for quicker and deeper debt relief to HIPCs which 'establish concrete and transparent linkages between debt reduction and a development strategy that focuses on reducing poverty' (OI 1998c: 2). This 'would offer an opportunity to transform HIPC into an effective strategy for supporting poverty reduction' (OI 1998c: 2).

The World Bank's poverty-focused debt relief framework: Oxfam's analysis

By October 1999, Oxfam was able to report 'fundamental and positive changes to the way the IMF and World Bank operate in poor countries' and that 'Bank/ Fund programmes cannot go ahead unless they are transparently designed to address poverty reduction' (OI 1999e: 1). These changes were regarded as 'a positive step forward, providing more relief to more countries, with a strong and transparent link to poverty reduction' (OI 1999e: 1). Oxfam compared seven key features of the original and enhanced HIPC frameworks and then summarised the link between conditionality and poverty reduction components as outlined in Table 5.1.

Throughout 2000, Oxfam maintained its critique of the enhanced HIPC, for '[t]o demand that governments in these countries spend more on debt servicing than on the basic health and education needs of their citizens is economically irrational, morally unacceptable and at variance with the HIPC Initiative's proclaimed goals of providing a poverty-focused debt relief framework' (OI

Table 5.1 Conditionality and poverty reduction

	HIPC	Enhanced HIPC
Conditionality	IMF is the gatekeeper on debt relief: progress under HIPC linked throughout to performance of IMF Enhanced Structural Adjustment Facility (ESAF) programme.	Major change, but unclear how this will work in practice. ESAF now renamed as the Poverty Reduction and Growth Facility (PRGF). Conditionality is now linked to achievements in implementation of a Poverty Reduction Strategy Paper (PRSP), which includes macro-economic stabilisation measures. While Oxfam agrees that macro-economic stability is required, it is unclear whether the IMF agrees to allow overall poverty reduction performance take precedence over traditional macro-economic policy reform measures.
Link to poverty reduction	Weak	Substantial. See above on 'Conditionality', but it is unclear just how this will work in practice. Nor does the enhanced HIPC framework provide enough finance, which in addition to increases in aid could enable many countries to meet the 2015 international development targets and thus properly address poverty reduction.

Source: OI 1999e: 11.

2000: 2). Oxfam concluded: 'The data derived from the Oxfam research strongly suggests that unsustainable debt will remain a formidable obstacle to poverty reduction efforts' (OI 2000: 3). Oxfam continued to press for reforms, which included a ceiling of 10 per cent of government revenues to debt servicing and immediate relief to countries which, in their interim PRSP, commit to introduce a 'poverty action fund' to be allocated to poverty reduction initiatives (OI 2000: 5).

In 2001, Oxfam argued that the 22 countries receiving HIPC relief were still spending an unacceptable proportion of their annual budgets on debt service compared to essential social services. In linking HIPC relief to the 2015 Millennium Development Goals (adopted at the United Nations Millennium Summit of September 2000), Oxfam urged the World Bank and IMF, at their spring meetings in 2001, to:

- endorse the principle that no HIPC country which is serious about the 2015 goals should be denied the resources required to achieve them;
- agree a new HIPC3, and ensure that future debt sustainability criteria are linked to the financing requirements of the 2015 goals in indebted countries;
- undertake an urgent debt sustainability analysis of all low-income countries, and widen the HIPC Initiative to more countries such as Haiti, Nigeria, Georgia or Bangladesh;
- agree to 100 per cent cancellation of IMF and World Bank debt for HIPC countries which have illustrated that they can use the resources to deliver poverty reduction and where a 10 per cent debt service to revenue ceiling is insufficient to release enough resources.

(OI 2001b: 3)

Its review of the development of policy linking HIPC relief with poverty reduction programmes in beneficiary countries demonstrated that the World Bank not only heard NGO calls for change but actively engaged NGOs in its deliberations and publicly acknowledged the role of NGOs in shaping its policy.

However, despite this shift in World Bank policy towards Oxfam's and other NGOs' policies, Oxfam has continued to argue for stronger reform. In 2005, while welcoming the G8's proposal to cancel all World Bank and IMF multi-lateral debt of all HIPC countries, Oxfam was still pressing the G8 to extend multilateral debt cancellation under the HIPC Initiative to other poor coun-tries, for this will 'need to go much further if all poor countries are to reach the Millennium Development Goals' (OI 2005: 1). Thus, Oxfam called for a broadening of the current debt relief initiative to more poor countries, to the extension of relief from multilateral debt to other multilateral creditors such as the Inter-American and Asian Development Bank, and to surety that debt relief would not be funded from existing aid budgets. It also argued that conditionality should be restricted to financial accountability and poverty reduction goals, agreed through open and transparent decision-making processes in which civil society and parliaments play an active role, thus freeing poor nations from what Oxfam regarded as harmful constraints.

Personal insights into HIPC history

The following sections of this chapter present findings from interviews with World Bank management and staff, Oxfam advocacy staff and senior manage-ment of Oxfam International affiliates, several senior G7 and OECD country politicians who played leading roles in the HIPC Initiative, staff of NGOs engaged in the HIPC Initiative, and journalists who had covered the matter. These semi-structured interviews sought to better understand World Bank policy formulation processes and the degree to which Oxfam's advocacy may have influenced policy. The interviews complement the preceding document-based research, providing informed insights into complex policy formulation processes, adding depth to the documented history of the HIPC Initiative and

Figure 5.1 Campaigning for debt relief

enabling detailed description of the process of linking debt relief with poverty reduction programmes in beneficiary countries.

One political leader and several World Bank and Oxfam staff members provided useful insights into the history of the HIPC Initiative. Historical background was principally provided by the World Bank president and Bank staff most directly involved in the early stages of the HIPC Initiative and in its evolution into the HIPC review and enhanced HIPC, the latter with a strengthened link with poverty reduction.

The World Bank president recounted the genesis of the HIPC Initiative:

> . . . there is quite a lot of revisionism about HIPC. The facts are that HIPC did not exist before I took that first trip to Africa [as World Bank president]. There may have been requests before that but they were going up against a brick wall because it was perceived that the international financial institutions would not move on the question of debt relief in any form and what was then only available was sort of round-robin activities through IDA where they had a way of essentially providing the monies on soft terms . . . But the issue of debt relief was something regarded as against the Almighty's rule and when I got back from Africa and had the good fortune to convince [Michel] Camdessus [the IMF managing director] that we should jointly try and put this [the HIPC debt relief proposal] up at the annual meeting. And that's exactly what happened. And I think it came

across to a surprised public [that the] two of us would present this programme which arose in my mind not from any comments from critics, but from my experience in Africa with a Muslim mullah who, on a trip to the Ivory Coast 12 days after I got to the Bank, described to me the problem that, if you put money in one pocket and took it out of the other, his parishioners would say to him he had nothing left. And it was done so simply and so straightforwardly that I thought he was right. And beating down on people and doing sort of round-robins was not sensible financing. I came back and expressed that to Michel [Camdessus] and he agreed. We had said the money should go for social purposes but then the civil society came in and basically said that now that we've got this vehicle it's not rich enough and it's not well positioned enough. I'm sure in the development of the idea [the CSOs] they kept everybody honest and they helped develop it.

A World Bank staff member, involved in the management of the HIPC Unit after its establishment, emphasised the priority given to debt relief by James Wolfensohn, who was appointed World Bank president in 1995. The importance of Wolfensohn's leadership in this regard, and of his dealings with and attitudes towards NGOs, was corroborated in Mallaby's account of his presidency of the Bank (Mallaby 2005). Prior to this appointment

World Bank policy had been driven by the IMF which divided countries into those with good and bad economic policy. IMF ideology was that countries with good economic policies would correct over time and so didn't need debt relief while countries with bad [economic] policy weren't entitled to support. This was rationalised into a policy of no debt relief at all.

Further 'intense grass root campaigning and Wolfensohn as President created a climate for change which gave rise to HIPC'. A critical point was mentioned 'around July 1997 when a draft HIPC paper showing the need for a $US12 billion debt relief trust fund was leaked to the *Financial Times* and *The Economist*', describing this as 'a killer to debt relief because of its scale'. It was also observed that 'in this critical period [from 1995] many organisations and coalitions created the political will for the Development Committee's instruction for the IMF and World Bank to come up with a workable debt relief programme'. The staff member saw Oxfam as having '. . . played an important and crucial role in following through with ideas to support the public pressure'.

Another World Bank staff member involved in the HIPC review in 1999 similarly described the background to HIPC, the setback as a result of the leak to the *Financial Times* and the opposition to HIPC from certain G7 creditor countries. For him, the German change of government in early 1999 and the holding of the G7 summit in Cologne in June of that year were a catalyst for reinvigorating HIPC and for the HIPC review's willingness to look more closely at the link between debt relief and poverty reduction. In the HIPC review, the World Bank consulted closely with NGOs, including Oxfam, which, he said, had 'extraordinarily good relevant ideas on . . . debt sustainability [and] certainly on the link to poverty reduction'. In describing the consultation process, he referred to the frequency of meetings with Oxfam WAO staff, and the close study of all papers submitted by NGOs, in which 'Oxfam's contribution was most easily identifiable'. Referring to the further development of the PRSP, he suggested that the participation of civil society in HIPC government poverty reduction plans was a very significant advance, coming very much from Oxfam.

A former EU development minister offered observations about World Bank shareholder political support for HIPC and the limitations of that support. She suggested what was needed to adopt and implement the World Bank's HIPC policy was a combination of the World Bank's management being intellectually convinced of its merits, and the shareholders' willingness to make it happen by financing the proposed relief. The importance of shareholder support was also emphasised in reference to the Jubilee 2000 campaign having 'prepared the ground . . . in the G7' and to the World Bank's inability to act decisively 'without having agreement of the G7, or having the money flowing from the G7'.

Historical insights were also gleaned from two WAO staff who had engaged in Oxfam's World Bank HIPC and HDW advocacy. To a former WAO advocacy director, the World Bank's initial interest in Oxfam's HDW proposals was

> for tactical reasons, because they [the World Bank] needed to convince sceptical governments such as the US that it was good to give debt relief so they needed to say it was going for poverty reduction and it was going in a targeted way and not for corruption.

He added:

> We wanted it because we believed it was the right thing to do. But because of the way the Bank had to sell it to the US and others, it made all of the other NGOs extraordinarily sceptical about it. They thought it was another form of conditionality.

Mention was made of Oxfam being attacked by Jubilee 2000 (of which Oxfam was a prominent member) and constituents of the 50 Years Is Enough coalition because 'They thought it was a cover for structural adjustment and conditionality.' One former WAO head also referred to the different motivations for supporting HIPC and the strong link with poverty reduction:

> there was another applicable dynamic which in the end prevailed [which] is that HIPC had to be seen to be delivered because it's clear [that] politicians who had agreed to it were under pressure for it to deliver and [this] became the most important pressure.

Oxfam's influence on HIPC policy

Two former managers of the World Bank HIPC Unit acknowledged the early influence of Oxfam's debt advocacy. One emphasised that Oxfam's role over the period 1995 to 1997 was 'unique among NGOs in that it built a high degree of credibility through a combination of building relationships around well researched and formulated proposals which were within the parameters of the possible'. The second 'worked extremely closely with Oxfam . . . as they were instrumental in helping to identify the technical basis for the enhanced HIPC debt sustainability parameters, delivery time and, most of all, the PRSP – how to link debt relief to poverty reduction'. In relation to Oxfam's general role related to the HIPC, the president of the World Bank observed:

> . . . the role of NGOs is critical in the sense that they monitor and they keep people's performance to a standard that either was originally set or at a higher standard. And . . . in that sense Oxfam was useful as . . . they were useful in pressing for additional funding . . . they were very useful on the direction. . . . on the additional funding, NGOs never fully understood that the issue was not an issue for the [World] Bank or the [International Monetary] Fund but it was an issue for our shareholders. But . . . the limitation was . . . the limitation of capital . . . [and] even today [on] debt relief it's a question of to what extent will governments put out money for debt relief as well as for development. So . . . it was great that they kept the pressure on. . . . I don't think the donor governments would have in the early years [been] ready to do it and . . . it gave the governments the opportunity to justify this sort of activity because it was for good solid social purposes. . . . that Oxfam was a very useful instrument in convincing the public and also an instrument of pressure to make sure that it was focused on the right issues, so I give them credit for that.

A former senior member of World Bank management, responsible for its work on poverty management issues, suggested Oxfam considerably influenced the

> design of the initial HIPC . . . looking at the issue of sustainability . . . [and] looking at strengthening the link with poverty. . . . In the end . . . we went in the direction the Oxfam research pushed us and other people's research pushed us.

In addition:

> Oxfam's position helped in solidifying or consolidating the support for the linkage in key stakeholders more than it did in pushing the linkage within the institution. . . . [In relation to the PRSP process with the strengthened link between debt relief and poverty reduction] it was quite useful because it reinforced the way we were trying to move in, but it was

another independent way of making that case more effectively with the stakeholders. I think Oxfam was quite effective in raising a general sense of support for the need to link this [poverty reduction and debt relief] tightly. . . . I thought that the timing gave focus and impetus to the HIPC revamping but it was driven to a large extent by the increasing voices from civil society direct to the institution and directly through their national shareholders. . . . In the revamping of HIPC, I would give them [Oxfam] more credit and say that they actually helped to push the institutions to do the revamping sooner than they might have done. I think that the timing of the review, the degree to which we were open on it, was partly due to the response to civil society and Oxfam.

A senior economist and director of development policy corroborated this:

Clearly Oxfam has influenced the Bank, so we would not be where we are on HIPC if it was not for Jubilee 2000, of which Oxfam was a part, and other campaigns. I think that Oxfam has probably pushed us faster than we might have otherwise gone on the whole mainstreaming of participation.

A leader of the 1999 HIPC review also observed: 'the debt sustainability ratios were too high, the six years' time-frame too long and the social spending link too weak. Oxfam helped generate consensus on how to address these.' He recalled that Oxfam's influence was, 'In particular, the contribution to achieving a more transparent and compelling linkage between debt relief and poverty reduction.' A sector manager within the Poverty Reduction Group (which is responsible for the PRSP process) observed:

NGOs arguing for the link between debt relief and poverty reduction were very influential. . . . Oxfam certainly played a role. . . . Because Oxfam made that link so strongly, it influenced the way in which a particular aspect of HIPC II was implemented.

A manager of the HIPC Unit agreed to be interviewed because '. . . it was important to recognise OI's role in shaping the policy agenda at least as it pertains to debt relief'. On the influence of Oxfam's HDW proposals he commented 'I think it has shaped it', but added 'You have to recognise that Oxfam was not the only one pushing for the HDW'. Additional comments about Oxfam's role are recorded in Box 5.2.

Oxfam's Human Development Window proposals

Oxfam's HDW proposals, to strengthen the link between increased HIPC debt relief and poverty reduction programmes, were credited as a significant factor in the September 1999 decision to introduce PRSPs. A leader of the 1999 HIPC review particularly referred to Oxfam's experience of Uganda as a precedent demonstrating how debt relief may be linked to poverty reduction programmes:

***Box 5.2*: Comments from World Bank personnel**

- 'Where we had the discussion was that you don't necessarily have the one-to-one relationship between the dollars saved from debt relief . . . and . . . education and health. NGOs played a very valuable role . . . [and] debt relief mobilised so many people that it was in a way the perfect campaign to redynamise the poverty debate to see that ultimately it ended up in the PRSP process. . . . [I]f you look in the 1999 review, it was very clear that people wanted to see that debt relief would have a closer link to the poverty debate.'
- 'In policy circles, very, very influential.'
- 'There is absolutely no question that Oxfam shaped the final outcome of the enhanced HIPC programme and especially PRSP. All one has to do is to look at their papers submitted for the HIPC review in spring 1999.'
- 'Overall, Oxfam (along with EURODAD) occupied a unique place, and can clearly be said to have changed the face of the [HIPC] Initiative both through its analysis and advocacy.'
- 'Oxfam matters when it counts. Their [Oxfam's] impact on HIPC, PRSP is evident . . . [and] they are very, very effective' (HIPC review member).
- 'I think both HIPC and debt relief and EFA are two areas where they [Oxfam] have . . . a pretty significant influence on policy direction, not just with the [World] Bank but even as importantly or more importantly with our shareholders. . . . [W]here they really made their mark was on debt relief' (Senior Communications Officer, Global Civil Society Team, External Affairs Department).

(Source: Interviews, 2002)

[Where] Oxfam's contribution was most easily identifiable is on the debt [relief] to poverty [reduction programmes]. . . . Kevin Watkins [of Oxfam GB] chaired a number of the sessions where this was driven by the Uganda poverty eradication action plan . . . where the money goes into a fund that has a bit more transparency. Basically Oxfam brokered the whole discussions.

About Oxfam's ability to influence World Bank policy, he asserted:

There is evidence, absolutely no question, Oxfam shaped the final outcome of the enhanced HIPC programme. I urge you to go to their contributions during the '99 review and then go to the modifications, then go to the website, the modifications paper, where [at] para. 16 the most detailed proposal . . . is Oxfam's and UNICEF's debt for development plan. . . . It's not whether they played a role but how they have done [their advocacy]. Basically you can go through the enhanced framework and see Oxfam's shaping of the way it finally came out.

Similarly, a former manager of the HIPC Unit commented: 'Oxfam's proposals, while not incorporated into HIPC II with the degree of direct ring-fenced nexus to debt relief proposed by Oxfam, were the catalyst for the linkage between poverty reduction programmes and debt relief under HIPC II and in PRSPs.' Additionally, a World Bank executive director agreed it was 'very valid' that the concept of PRSPs had its genesis in Oxfam's linking HIPC relief with poverty reduction programmes.

Others were either more equivocal or disagreed with this proposition. An External Affairs staff member acknowledged Oxfam was 'influential on the debt link', but had difficulty in assessing the degree of its influence. Several dismissed Oxfam's influence on the debt relief/poverty linkage and introduction of PRSPs. Two HIPC Unit managers referred to what they believed to be Oxfam's impractical and inappropriate direct nexus between debt relief and poverty reduction programmes. One World Bank executive director described Oxfam's HDW proposals as 'total nonsense . . . but interesting to read' and his 'worry with Oxfam's proposition . . . of linking debt relief very closely with windows of health and education'. He criticised the Oxfam approach as

> a mistake because the PRSP should work through a total poverty strategy of a country and, if you focus too narrowly from the start on two important composite parts of that strategy, then you may be undermining the strategy as a whole . . . and we were not happy that Oxfam had appeared to have fallen into that same trap.

Another criticism of Oxfam's HDW proposals was that they emphasised 'preconditionality in debt relief and so . . . [indicated] that Oxfam wants structural adjustment'.

Despite these views, the UK Chancellor of the Exchequer's Office acknowledged Oxfam's distinctive influence as arising from Oxfam being the 'first NGO in the whole debt campaign who made a stance on linking debt relief to a tangible outcome . . . bridging the gap on conditionality issues and . . . therefore part of linking debt relief to creating PRSPs'. The Chancellor of the Exchequer's Office also believed that Oxfam 'had an impact on moving [the] UK government's position on HIPC ratios (especially fiscal) and therefore [the decisions taken by the G7 meeting in] Cologne. This had an impact on linkage to PRSPs.'

A former EU development minister was more expansive, referring to Oxfam's very early involvement in the HIPC Initiative:

> [Oxfam was] ahead of the pack in raising this issue and personally I believe that, in these first meetings with the Board of the World Bank, the idea was

shaped on what the debt campaign should actually be about. Oxfam was . . . convinced at the time that it should not be unconditional debt relief to everybody and their grandmother. There should be an explicit link between money that was saved . . . and spending . . . in the social sectors as one way to make the whole initiative marketable to those in power and credible. That idea of who is going to benefit from this initiative came up in the first conversations, as it was their idea of who is going to pay for debt relief, and how important it was to get donor countries to put money on the table to enable the international institutions to write off some of the debt.

So Oxfam, right from the beginning, helped to shape the link between debt reduction and social sector spending, and really they educated the campaign because Jubilee, at the beginning, was really the biblical idea of cancelling all debt for everybody and their grandmother in the year 2000, which would never have been influential as such. . . . Oxfam had prepared the ground on what kind of ideas are marketable or not.

Further, in relation to the debt relief programmes implemented under the HIPC Initiative, to the same former minister 'Oxfam prepared the ground for both shareholders and management of the Bank to start thinking and recognising that this is a serious problem. . . . Oxfam has been very helpful in getting their partners in the South to get on board.'

Oxfam senior management in the WAO and the affiliate advocacy staff commented positively on the influence of Oxfam's HDW proposals on relevant World Bank policy. However, Oxfam staff were less specific, commenting generally that Oxfam's HDW proposals had a positive influence on the World Bank's strengthened linkage between HIPC relief and poverty reduction programmes in beneficiary countries. Although such 'in-house' support might be expected, it is nevertheless illustrative of a prevailing, although unevaluated, Oxfam view that it is effective in its World Bank-oriented advocacy.

A different perspective on the degree of Oxfam influence was provided by the executive director of Oxfam Canada, who, while rating Oxfam's influence as 'among the top' of NGOs, expressed the view: 'I still would not rate it as highly influential compared to say business groups or Northern governors of the [World] Bank.' Thus, there is recognition within Oxfam that it is just one influence among many, and other influences may limit the extent of that exerted by Oxfam.

Other NGOs' views about Oxfam's influence were generally more equivocal, especially in attributing that influence to the HIPC policy change. A Bank Information Centre staff member, while wondering about 'how positive change on something like external debt really occurs', suggested 'it is clear that Oxfam played an important role in helping move the discussion forward within the [World Bank] institution'. The US representative of ActionAid commented on the general difficulty of attribution of influence to any one organisation:

there is an identification of Oxfam with the HIPC process and the initial breakthrough for debt relief. . . . They are in the right group of people. . . .

Oxfam has done a lot of work in terms of research and has had the ability to influence policy.

Another NGO executive observed:

It is hard to distinguish Oxfam's particular role from the broader role of NGOs. It is easier for me to see influence of the NGO community of which I see Oxfam as a leading member. There is no question that the NGO community as whole has a strong influence on [World] Bank policies and [US] Treasury policies on HIPC and on PRSPs.

Three of four journalists who had reported on the increased World Bank emphasis between debt relief and poverty spoke positively about Oxfam's direct influence on policy development. An Agence France-Presse (AFP) journalist indicated the strongest Oxfam influence. He referred to the World Bank's perception of Oxfam as 'the only one [NGO] the [World Bank] are ready to work with. . . . [Oxfam] is positioned between the optimistic and pessimistic extremes of the HIPC debate'. Then:

[In relation to HIPC] Oxfam was . . . saying it is a programme that may work but to achieve anything you need to be careful about how the [debt] relief is structured. Oxfam came up with a report saying it should work. . . . Until then Oxfam was the only NGO working with the World Bank and IMF [and] showing that HIPC was not working. This caused a big shock at the [World] Bank. . . . Oxfam was the one NGO making the [World] Bank and IMF see that HIPC was not working and it would have taken much longer to find out if it were not for Oxfam.

In describing Oxfam's approach to seeking workable proposals, he added: 'Every agreement is a compromise and Oxfam is very good at seeing that compromise [is needed] from the beginning and knowing how to get there.' The *Washington Post* journalist indicated that he thought the World Bank was already heading in the same general direction as Oxfam's HDW proposals. After expressing the view that 'I don't see much difference between Oxfam's position and the [World] Bank's position', he referred to Oxfam as

an ally, not an adversary, of the [World] Bank. . . . [Oxfam] added energy to the [World] Bank in that particular area, which means that the [World] Bank would not have been quite so effective by itself . . . so Oxfam's approach was very effective.

The Canadian *Globe and Mail* journalist similarly observed that 'Oxfam was the source of a great many ideas that were debated by World Bank and IMF personnel' and then commented:

the [World] Bank and Oxfam seem to be on exactly the same wavelength
. . . except that Oxfam is more outspoken about the need for governments
to put more money on the table. . . . The [World] Bank's recommendation
to change the emphasis of the HIPC Initiative seems to be in line with what
Oxfam has proposed.

However, a *Financial Times* journalist was less positive, referring to NGO HIPC
advocacy more generally and to Oxfam in particular thus:

> One thing it [Oxfam] has not done is to reopen the fundamental questions
> of how much debt relief should be given and how much should be linked
> to the Millennium Development Goals. . . . No NGO has made any serious
> inroad in either the [World] Bank or the [International Monetary] Fund or
> indeed to any one of the shareholder countries.

Once again, in the world of journalism there was a diversity of perceptions
about the exact role of Oxfam and other NGOs in affecting World Bank policy
directions.

Oxfam's World Bank shareholder-oriented advocacy

While it is evident that Oxfam had direct influence on the formulation of World
Bank policy, it was generally perceived to have had less influence on the Bank's
shareholders. Two interview participants referred to the importance of shareholder policy (and those shareholders' executive directors) in framing World
Bank policy, but neither directly acknowledged any Oxfam influence on
shareholder policy. One executive director, while affirming the roles of her
government's finance minister and World Bank governor, was silent on any
Oxfam influence on her government's position, instead suggesting that 'Oxfam's
involvement on HIPC sustainability and the engagement of strong shareholders
really led to a review of the sustainability of the HIPC programme'. Another was
a little more direct, referring to Oxfam having helped in 'solidifying or
consolidating support for that [poverty and debt relief linkage] among key
stakeholders' and mentioning that '[Oxfam in] the UK and US in particular
were very effective in what we were trying to do'. He also referred to the HIPC
review being 'driven to a large extent by the increasing voices from civil society
. . . through their national shareholders [pressing] for the need to rethink the
previous thing'.

Oxfam staff were generally more specific about influences on affiliate
shareholder governments, seeing influence on the US government as critical. Of
Oxfam America's role, in conjunction with Jubilee 2000 and Oxfam International, the executive director of Oxfam America observed:

> Oxfam America was . . . inside the [World] Bank and the US Treasury
> Department and the US Congress doing all sorts of lobbying on the HIPC.

... [A]t the end of the day the debt relief deal was done through a high degree of coordination between the OI and OA staff, delivering the US Treasury and also delivering the Republican Party leadership on the Hill. ... [W]e [OA] were on a few organisations lobbying the Republican side of the House to get this work done.

Similar comments were made by staff from the Australian and Canadian Oxfams in relation to their governments' policies on the HDW aspects of the HIPC Initiative. Surprisingly, in light of the UK Chancellor of the Exchequer's Office comments on Oxfam's degree of HIPC and related HDW influence, the Oxfam Great Britain participant made no direct reference to having influenced the UK government's policy position.

Just one of the NGO participants commented on the influence of World Bank shareholders in its policy formulation and on HIPC (particularly in the US context): 'HIPC would not have happened if the G7 had not agreed to it no matter what the World Bank wanted to do. In influencing G7 membership, Oxfam was one of the leaders.' British newspaper journalists referred to Oxfam's influence on the relevant policy of the UK government as one of the significant World Bank shareholders, positively linking this influence to the UK Jubilee 2000 campaign and Oxfam's part in it. In reference to the UK Chancellor's commitment to debt relief, the *Guardian* journalist spoke of 'A marked shift of tone . . . [which] was the result of lobbying by NGOs such as Oxfam' and to 'Oxfam doing the detailed thinking which the [World] Bank obviously picked up. . . . [Y]ou can see it reflected in their papers'. Oxfam's involvement in Jubilee 2000 and the fact that tens of thousands sometimes stood outside G7 meetings also 'made it clear to political leaders like [Gordon] Brown [the UK Chancellor of the Exchequer] . . . that there was a political payoff for being seen to move on this issue'. Observations by a *Financial Times* journalist were similar:

> Obviously Jubilee 2000 was a stand-alone thing, but backed by Oxfam, and Oxfam provided a lot of the intellectual clout behind it. [Jubilee 2000] was primarily effective because it was directed at the shareholders (Gordon Brown, etc.) so it became a big issue within the country and clearly did influence [World] Bank policy. It is not clear to me that directly lobbying Bank staff would get the same effect.

The influence of Jubilee 2000 and other NGOs

The greatest difficulty in assessing the impact of NGO advocacy was disentangling Oxfam's influence from that of others. World Bank staff referred to either the debt relief campaign and the difficulty of attribution of influence to any one advocate, or to Jubilee 2000 and Oxfam's participation in it. A number of participants specifically distinguished Oxfam's membership of Jubilee 2000 from its own debt relief and HDW advocacy, and several acknowledged the political pressure for debt relief created by the wider campaign in which Jubilee 2000 was prominent. The roles of Jubilee 2000 and Oxfam, and the difficulties of attribution, were recognised in a range of representative comments, including:

> Certainly, we should not underestimate the importance of the international pressure in part driven by Oxfam. Once the various forces came together – Jubilee 2000, Oxfam work on debt, etc. – they generated a global momentum that is difficult to attribute to any one agency.
> (World Bank Education Team executive)

> Jubilee was able to galvanise a movement in the way the Oxfams really couldn't. . . . Oxfam was seen as the NGO with the best knowledge of the issues, but no matter how excellent that knowledge is and how good those papers are . . . G7 leaders need to get a sense that the people in that country are angry. That's what Jubilee was able to do.

Oxfam affiliate staff acknowledged the role of other NGOs, referring to Jubilee 2000 as having contributed to political pressure, which influenced their governments' support for the HIPC Initiative and later for an increased emphasis on the linkage between debt relief and poverty reduction. Representative comments acknowledged that Jubilee created 'political space' but also that Oxfam 'managed to be inside the [World] Bank working on the technical details with regard to the HDW'.

Figure 5.2 Oxfam campaigners, part of the Jubilee 2000 coalition

Representatives of NGOs variously recognised the difficulty of attribution of influences as well as of Oxfam's role in supporting Jubilee 2000 from the perspective of both borrowing HIPCs and donor/World Bank shareholder advocacy perspectives. One commented: 'I'm not sure you even would have got there if it was not for the work that Oxfam was putting forward and doing.' However, the range of comments suggest that Jubilee 2000's campaign was instrumental in creating the political support for action and therefore World Bank and shareholder support for Oxfam's HDW proposals.

Internal dynamics of the World Bank HDW proposals

It was not only a combination of the wider campaign for debt relief and proposals from Oxfam for a stronger link between HIPC relief and poverty reduction which resulted in the HIPC review recommendations for that strengthened link and for the introduction of PRSPs. There were also forces at work within the World Bank pressing for such policy change. Several interviewees referred to the influence of the World Bank president, James Wolfensohn, who, since his appointment in 1995, had emphasised the World Bank's poverty reduction role and, in so doing, had empowered progressive and reformist staff members. Only one executive director suggested that World Bank processes, by themselves, would have brought about change, with NGOs, including Oxfam, having little influence.

Oxfam's influence on the World Bank's PRSP policy

Several staff from the World Bank and Oxfam specifically attributed the World Bank's PRSP policy to Oxfam's HDW proposals. Another raised NGO criticism of Oxfam's pressing for a strengthened linkage with poverty reduction, and the interaction between the World Bank and US government, as indicating a new form of conditionality for countries obtaining debt relief. A former director of Oxfam's WAO observed that 'the things that Oxfam was arguing as part of the HIPC process were important in having national poverty strategies owned by countries' and 'the biggest contribution . . . we made [was] making the link between debt and poverty reduction. The . . . PRSP . . . in a way came from this.' The Oxfam America director of policy linked PRSPs directly to a combination of Oxfam's HDW advocacy and US government policy by observing that 'the PRSP and HIPC Initiative grew out of Oxfam proposals, although actual formulation was a US Treasury product, and there was an OA lobbyist [who] worked very closely with the US Treasury'. She added:

> Oxfam is one of the catalysts of PRSPs. [It] provided a way out of a conundrum and gave a new twist on how to work on debt, and [the US] Treasury took that idea and ran with it. [It is an] example of [a] credible, positive Oxfam proposal.

The executive director of Oxfam America was, however, less sanguine about Oxfam's HDW proposals being a major influence in the formulation of World Bank PRSP policy, commenting: 'We have also played a role in the policy work in terms of the way the PRSP has gotten up and [been] implemented, although we are not alone. Attribution on that is doubtful'. Although two NGO participants linked the introduction of PRSPs with the HIPC Initiative, neither had sufficient knowledge to identify any direct Oxfam influence. Thus, representatives of other NGOs were less likely to acknowledge Oxfam's influence but, not being at the receiving end of Oxfam advocacy, were perhaps less well placed to make judgements in this respect.

Again, the *Financial Times* and *Guardian* journalists made the connection (although somewhat equivocally) between Oxfam's HDW advocacy and the introduction of PRSPs. Referring to Oxfam's linking of debt relief and poverty reduction progress and Jubilee 2000's call for total and unconditional relief as creating 'tension between the approach taken by OI [and] . . . others', the *Financial Times* journalist observed: 'Oxfam has to be taken seriously, has to have some credibility. How much effect that had on PRSP processes . . . is hard to say.'

Oxfam's advocacy style

While opinion may have been divided about the influence of Oxfam's HDW advocacy, there was greater consensus about the advocacy 'style' Oxfam employed. Several participants referred to that style being characterised by a combination of building relationships of trust and a sense of partnership, while maintaining a degree of independence which enabled Oxfam to be fiercely critical. One World Bank executive described this as

> entering into a dialogue. . . . [T]hey made critical remarks. . . . [This] is, I think, necessary. I am one at the [World] Bank who thinks we cannot do without it. . . . In that process [Oxfam] has been quite good in being positively engaged, then again criticising. . . . I think what was probably the main thing was that Oxfam commanded credibility. They were critical but they also provided good analytical background to it in their reports. [We] may not agree but that does not matter. Oxfam were not only critical but also provided an alternative on the table. That was very, very powerful.

Others referred to Oxfam's style as pragmatic and solution-oriented, using the media effectively in pursuit of its goals. One World Bank participant, comparing Oxfam's approach to that of Jubilee 2000, remarked:

> in terms of at the table, contributing ideas, working on policy, pushing and pulling and ideas on compromises, they [Oxfam] were crucial. . . . [Y]ou could think of it like this – Ann Pettifer[7] and Jubilee 2000 would come in and knock you on the head with a brick, and then Oxfam would walk in and

say OK, Ann wants this, [which] is not what is going to happen. This is what the current status quo is. What we all want to do is make the link between debt reduction and poverty reduction real. What's the best way to do it?

The World Bank's view of Oxfam's advocacy style is perhaps best encapsulated in the following observation from a 1999 HIPC review leader:

> [Oxfam] is much more influential because they tap into [and] acknowledge that there's a lot of discussion that takes place here about how we do this work . . . better, and they contribute to the thinking process. But . . . that doesn't mean they're not tough, or doesn't mean they don't slam us when we deserve it, and we deserve it a lot, so I think they are very, very effective.

World Bank staff positively referred to the quality of Oxfam's ideas and its research and analysis, specifically in relation to HDW proposals. Participants spoke of Oxfam's 'extraordinarily good, relevant ideas', its 'radical yet achievable positions in relation to debt sustainability', its substantive arguments and 'presentations which were quite compelling so you had to figure out how . . . you deal with them!' Two referred to Oxfam's research on African countries, which were based on having access to the Ugandan government (which increased the credibility of its proposals) or on original case study research which demonstrated the unsustainability of Tanzania's debt service burdens. According to an executive director, this 'was the first factual case study that had been done . . . on the assumptions and on sustainability of the programme. Even with all the [World] Bank's resources.' After referring to Jubilee 2000 people as 'crazy, nuts . . . they don't know what real life is about', another executive director who was generally sceptical about the value of NGO advocacy remarked that 'Oxfam comes much closer – they may often be right'. The other substantive observation from a senior officer now with the IMF was that

> What sets aside Oxfam's effectiveness compared to others is that it is not as single-issue as others tended to be. I always found that their analyses generally take a broader view and therefore are more conscious of trade-offs than some of the other pieces I get. Also analytically they are generally of good quality. They are not advocacy pieces based on obvious flaws – selective use of data, selective presentation of the facts – so that you can readily engage. It is not a polemical argument. . . . [T]hat is a really important thing to retain. . . . [T]hey clearly are there from the perspective of civil societies. They pick their issues from that perspective but then they bring to it the rigour of analysis that enables them to be credible partners in the policy-making discussion. There are not many people who do that.

These comments demonstrate the importance of NGO advocacy being clearly based on sound research.

Rose-Breasted Grosbeak

2014

forever
U S A

P1111

Staff from Oxfam also commented on its advocacy style. The first WAO advocacy director attributed Oxfam's influence on World Bank policy to an understanding of World Bank politics and the tactics required to achieve influence: '[The] contribution we made [was because] we understood at the time how to play it. We played [the] lobbying and media game very cleverly because we understood the politics and that is what Oxfam does well and understands.' Oxfam Australia's[8] executive director also referred to this 'insider' approach and understanding of World Bank and shareholder politics, expressing the view:

> Oxfam were able to move proposals forward because of the empathy and familiarity between the [World] Bank's staff and the president and his understanding of the [World] Bank's pressures and the national pressures. It was the difference from an insider and outsider perspective.

The Oxfam America executive director's view noted a distinguishing feature of Oxfam's advocacy – a willingness to engage with unsympathetic US Republican politicians who controlled the US Congress and who were therefore influential in the formulation of US policy. Such policy is frequently implemented because the US is the largest and most influential World Bank shareholder. The Oxfam America policy director attributed Oxfam's influence to a combination of 'very seminal research, an idea of how to wiggle out of a dead end, important lobbying and very good press work'.

Research participants were requested to rate Oxfam's effectiveness and, relative to other factors at work, its influence as an advocate in shaping World Bank policy in relation to proposals to directly link increased HIPC debt relief to poverty reduction programmes in beneficiary countries. Among the World Bank personnel and politicians interviewed, 10 of the 11 rated the effectiveness or influence of Oxfam's HDW advocacy as 'high' (the president's rating) or 'very high'. Three of the four NGO staff and journalists similarly rated Oxfam's HDW advocacy effectiveness. Interestingly, Oxfam advocacy staff and executive directors were somewhat more modest in their assessment, measuring policy outcomes against ambitious and broader debt relief advocacy objectives, of which the HDW achievements were just part.

Conclusion

Oxfam's HDW proposals and supporting advocacy had a clear and material influence on relevant World Bank policy. From 1997, when Oxfam began pressing for a strengthened link between HIPC relief and poverty reduction programmes in beneficiary countries, until the World Bank's policy decision following completion of the HIPC Phase II review in 1999, there was significant movement towards Oxfam's HDW proposals. Oxfam's influence on the various areas of World Bank decision making was acknowledged in the formal World Bank papers, by persons in all categories of those interviewed (including the

World Bank president and especially staff directly responsible for the HIPC programme and review), and in the overall ratings of Oxfam's HDW advocacy, as measured by its influence on policy outcomes.

Oxfam's influence was also reflected in the World Bank's introduction of PRSPs. These are conceptually consistent with its calls for a poverty reduction framework as a prerequisite for HIPC relief. They also provide for in-country civil society participation, something for which Oxfam and others had argued. A number of interview participants directly attributed the PRSP policy to the World Bank's acceptance of a strengthened link between HIPC relief and poverty reduction, an area where Oxfam's influence was clearly evident.

Despite this strengthened link between HIPC relief and poverty reduction, Oxfam's position, reflected in its April 2001 paper, was that World Bank policy still left poor countries with unacceptably high debt service burdens compared to their capacity to fund essential social services (OI 2001b). That same paper reflected Oxfam's view that further measures were required to achieve levels of debt sustainability compatible with realisation of the 2015 Millennium Development Goals, a position which Oxfam had maintained through to late 2005 (OI 2001b; 2005). So, despite the success of its HDW advocacy efforts to more closely link debt relief with poverty reduction programmes in beneficiary countries, Oxfam's broader HIPC advocacy objectives have not been fully achieved. Nonetheless, its advocacy has enabled and produced subtle yet significant policy changes in the developing world's most important bank.

Part III

A hesitant courtship

Engaging the corporate sector

6 Confrontation, cooperation and co-optation

NGO advocacy and corporations

John Sayer

I am here because there is poverty in the Third World while these companies get richer and richer.

(Street demonstrator at World Economic Forum,
Davos, 2001, BBC TV News)

As we pursue our strategies worldwide, we accept social and environmental responsibility, including the promotion of a sustainable global economy and recognition of our accountability to the economies, environments, and communities where we do business around the world.

(General Motors CEO, John Smith, in GM's 1996
Environmental Annual Report)

A hesitant courtship: NGOs and corporations

Corporations and NGOs are both searching for the right terms of engagement in the constantly changing economic and political environment. This process only started to gain significance in NGO policy making in the 1990s, but is now seen by some as one of the most important dynamics in the changing power relations of a globalised world. NGOs, and particularly the development NGOs which are the focus of this book, are still at work clarifying their underlying objectives for interaction with the corporate sector, defining their basic policies and shaping consistent approaches and procedures.

The influence and power of businesses, particularly large transnational companies, has grown in recent decades. By 2000, the top 200 corporations' combined sales were bigger than the combined economies of all the nations in the world except for the biggest ten. The top 200 corporations' combined sales were 18 times the size of the combined annual income of the 1.2 billion people (24 per cent of the total world population) living in 'severe' poverty (Anderson and Cavanagh 2000: 1). This awesome economic might, heedless of national boundaries, has given corporations immense influence with governments and multilateral institutions in relation to macro-economic debates about structural adjustment, trade, debt, investment and finance. Thus, 'Corporations have emerged as the dominant governance institutions on the planet, with the largest among them reaching into virtually every country of the world and exceeding most governments in size and power' (Korten 1995: 53–5). In this era of globalisation, privatisation, deregulation and neoliberal economic ideology, corporate power has also grown in relation to that of national governments, which have declining influence over trade, investment and finance policy. One study of corporate activity in a globalised world asked:

> Who, today, can effectively regulate an oil company active in 160 countries? a clothing manufacturer with factories in 50 countries? ... What governments would even try? Instead they seem intent on eliminating what vestigial regulation they do practice in order to be more attractive to foreign investors who provide their constituents with jobs. And would the old style of regulation even work under the new order?
>
> (Schwartz and Gibb 1999: 22)

In addition to political and economic influence, companies increasingly control technological advances and information and will therefore influence the shape of future development and the distribution of prosperity (Korten 1995).

Compelling evidence shows that economic growth is a pre-condition for sustainable social development and poverty alleviation, despite the fact that growth on its own does not ensure either sustainability or sufficiently equitable distribution to reduce poverty (UNDP 1996; Hanmer *et al.* 2000; World Bank Group 2000). The corporate sector is clearly the main engine of that growth, playing a key role in the creation of jobs, the generation of tax revenue, the earning of foreign exchange, the generation of finance, the achievement of access to new markets, and the transfer of technology and administrative skills (Oxfam *et al.* 2002).

At the same time, corporations have the potential to create immense problems for poor people. Profit maximisation and competition motivate corporate activity; unregulated, this can have a negative effect on natural resources, on the environment, on local prices and on jobs. Large foreign corporations can destroy smaller local business, monopolise markets and reduce the access that others have to productive opportunities. All of these are particularly vital to the livelihoods of poor people. The corporate sector has little orientation towards the creation of equity, and no obvious incentive to ensure that its goods and services reach all people (Korten 1995; Oxfam *et al.* 2002).

Global integration has spread and the market economy has become ubiquitous, yet figures indicate that the gap between the richest and the poorest people as well as the richest and poorest nations has grown throughout the last few decades of liberalisation (Korten 1995). The gap between the richest tenth of humanity and the poorest tenth grew from a ratio of 52:1 in 1970 (UNDP 2001: 20) to 103:1 in 2005 (UNDP 2005: 38). Two and a half billion people live on less than two dollars a day and overall progress in reducing this number has fallen (UNDP 2005: 34). These facts alone challenge more simplistic theories that freer markets, liberalised trade and free-flowing investment will benefit all that they touch and will, unfettered, bring more prosperous and indeed more democratic societies.

In the last decade, foreign direct investment and equity investment in most developing countries has far outstripped official development assistance, often referred to as foreign aid. Aid levels continue to decline, and many donor governments and multilateral development banks are proposing that creating an 'enabling environment' for the corporate sector and providing incentives for investment are more effective ways of promoting development than direct aid (ADB 2001: 22–5). Companies are increasingly involved in providing many of the services and much of the infrastructure formerly considered the responsibility of the state or local government such as water supply, education, transport, communications, education and health care. Thus, the private sector is engaged in vital components of the development process, which brings them into direct working relations with poor people and needy communities (Eade and Sayer 2006). Therefore, 'In moving towards a sustainable global economy, we will depend on the motivation, ambitions and performance of corporations – and on the restructuring of the markets which they serve' (Elkington 1997: 101). While corporate activity can make immense contributions to the development process, poor corporate practices can harm the lives of poor people who are already vulnerable.

As the power of corporations grows, so do demands for them to behave with a greater sense of environmental and social responsibility, by reducing the problems their operations may create and by demonstrating a positive role in the countries and communities where they operate. Until the 1990s, corporate social responsibility was thought of as paying tax, delivering a quality product, providing employment and abiding by the law. This has changed. By 2000, for example, 80 per cent of Australians believed large companies should be actively involved in environmental and human rights issues, and 70 per cent that chief executives should take an active stance on major social issues (Burbury 2000). Many leading international companies have accepted the validity of some criticism, acknowledging an implicit social responsibility. Corporate codes of conduct and social and environmental annual reports have become widespread. Some 45 per cent of the top 200 US corporations produce regular sustainability reports (KPMG 2002). In some cases these are accompanied by broader efforts to re-orient the core culture of companies towards greater social and environmental responsibility (Covey and Brown 2001).

In the face of economic liberalisation and globalisation, the capacity and willingness of governments to regulate corporate activity has diminished; they have reduced their function as corporate governors and regulators, and lessened their role as mediators between the demands of capital and the demands of society as a whole. Into this space has stepped civil society, expressed in terms of campaigners, protesters and consumer activists pressuring companies on environmental, social and labour issues. Simply in terms of size, civil society has grown rapidly in the past ten years. The United Nations estimated that the number of NGOs grew from about 29,000 in 1993 to more than 100,000 in 1999, and this growth was seen in both developed and developing countries (Edwards and Hulme 1996: 1). NGOs have been foremost among critics of corporations, highlighting corporate malpractice, campaigning publicly against those with poor records on the environment, labour and social impacts, and lobbying governments and international meetings in proposing more responsible paths for the corporate sector (Bendell 2000). Thus, 'Overwhelmingly these groups want two things from business – greater accountability and greater community involvement' (*Fortune* 1999: 3). Many believe this new role is building into a new form of 'civil regulation' or 'social licensing' of economic activity. International NGOs are thus critical players in the creation of these new global norms and this emerging informal governance.

NGOs have traditionally had two primary but quite antithetical areas of interaction with corporations: that of beneficiaries from corporate donations to charity, and that of hostile critics of corporate activities. A third area of contact is growing in importance: cooperation as project partners and consultants. Corporations are increasingly seeking NGO advice and cooperation, not only in joint work on social and environmental projects in the community, but also in ways to make their own core business practices more socially and environmentally acceptable:

> To achieve outstanding triple bottom line performance, new types of economic, social and environmental partnership are needed. Long-standing enemies must shift from mutual subversion to new forms of symbiosis.
>
> [. . .]
>
> But it is still far from clear how many of today's campaigning organisations will successfully make the transition to the news ways of operating.
>
> (Elkington 1997: 220, 223)

The engagement of these two powerful sectors of global society – companies and NGOs – on questions of equitable and sustainable development is of profound significance for the development process. The 1990s witnessed an unprecedented surge in interest and activity in interaction between firms and non-governmental organisations, with an impressive number of initiatives currently in operation across the globe (Crane 2000). This has received scant attention in the literature and, at discussions about equitable growth and poverty reduction, often 'the world of business lurks in the shadows, acknowledged uneasily like a tattooed man at a tea party' (Sayer 2005: 251). More substantial reflection and analysis of these relationships and their role in influencing the success or otherwise of programmes designed to promote sustainable development are overdue (Crane 2000). This chapter therefore explores aspects of the evolving connections between international development agencies and the corporate sector. It details the forces at work propelling NGOs and companies into newer and closer relationships in pursuit of their different missions and the different ways NGOs and companies interact, and examines relevant policy issues.

NGO engagement with the corporate sector

Interaction between transnational corporations (TNCs) and NGOs has occurred since the 1970s, but much early contact was characterised by ignorance on both sides, an absence of cooperation, distrust, aggression and confrontation (Enderle and Peters 1998; Crane 2000; CDCAC 2002). Campaigning organisations are, somewhat ironically, usually risk averse and find it more comfortable to tell others what they should not be doing, rather than offering, and taking responsibility for, possible solutions (Elkington 1997).

Despite this, in response to criticisms, corporations have become aware of the need to acknowledge new responsibilities, and a number are working in partnership with international NGOs on long-term development and social and conservation projects. Regardless of whether the motivation is one of cynical public relations or a genuine desire to address the issues, companies are approaching NGOs and other parts of civil society seeking advice, ideas and cooperation in policy formation, standard setting, promotion of best practice, monitoring and evaluation, and implementation of programmes and policies. Because much of the pressure for improvement emanated from NGOs, it is increasingly difficult for them to refuse to get involved when companies acknowledge these criticisms and seek help with change. Each side is investing knowledge, staff and finance to achieve agreed goals. Increasingly, stakeholders

> . . . see a convergence in rights and expectations for business and civil society; business leaders increasingly recognize an implicit social responsibility of businesses and the possibility that civil society actors can help business actors meet their obligations as good citizens. . . . Similarly, civil society organizations experiencing the costs and limitations of confrontational influence strategies see the potential for harnessing market forces in the service of their social goals.
>
> (Covey and Brown 2001: 20)

Even NGOs antagonistic to the corporate sector accept that engagement will grow. A 1998 survey of Northern and Southern NGOs found 82 per cent of surveyed NGOs believed that cooperative relationships with TNCs were possible. Although only 12 per cent then had such relations, a further 55 per cent believed they would develop in future (Enderle and Peters 1998: 23–4). A 1997 survey of development and environment groups found that more than 85 per cent also believed partnerships with companies would increase in the following years (Elkington 1997: 226–8). Civic groups have the potential to be 'a unifying influence, a bulwark against the abuse of power, and a practical vehicle for the transformation of values and behaviour'. To achieve this, though, they need to 'work much harder with business and government to make the social virtues less the property of one section of society, and more the defining characteristics of society as a whole' (Edwards 1999: 161).

Environmental groups have led the way in developing strategies and policies for interaction with companies, both cooperative and critical, partly because there are more causal links between business and the environment than between business and social justice (Heap 2000). Development NGOs still need to convince both society and corporations of the importance of addressing global social justice issues before they can shift from problem-focused to solution-focused advocacy (Heap 2000). However, businesses lack good social performance indicators; the state of these is equivalent to the state of environmental performance measures of 15 years ago (Henderson 2000). Moreover, 'While some aspects of environmental reporting lend themselves to the use of quantified indicators (emission per unit of product, for example), many social issues require a more qualitative approach' (Bendell and Lake 2000: 228). There is greater consensus on environmental issues, which can employ universal scientific principles, whereas social responsibility raises complex questions of cultural and ethical relativism, and the balance between social, political and economic rights. Social responsibility has, until recently, been largely the domain of governments and organisations such as trade unions, but increasingly social accountability is seen as something that extends beyond the boundaries of one country and is increasingly viewed as the responsibility not only of governments but also of corporations (Blowfield 1999).

NGOs have, in the past few years, begun to transform a rather ad hoc and reactive approach to relations with companies into more coherent and organisation-wide strategies. The relationship with the corporate sector can be complex, where NGOs are both beneficiaries and critics of corporations. As some NGOs have moved towards a more thorough analysis of the economic causes of poverty they have questioned the impact of some fundamentals of corporate activity such as foreign investment, financial flows and international trade. NGOs are relative latecomers to work on corporate accountability, having only recently begun to pay attention to market-based mechanisms (Bendell and Lake 2000). Thus, 'NGDOs [non-government development organisations] are simply ten years behind ENGOs [environmental NGOs] in terms of organisational capacity and outlook' (Heap 2000: 95).

New motivations for NGO engagement with companies

We, the people of the world will mobilise the forces of transnational civil society behind a widely shared agenda that bonds our many social movements in pursuit of just, sustainable and participatory human societies. In so doing we are forging our own instruments and processes for redefining the nature and meaning of human progress and for transforming those institutions that no longer respond to our needs.

(People's Earth Declaration from the International NGO Forum
of the Earth Summit, 1992 – NGO Forum 1992)

The role or influence of governments in global economic systems has been in decline, following increased domestic deregulation, greater emphasis on voluntarism, freer world trade and the growth in size of transnational corporations themselves (Nelson 1996; Waddell 1997; Rodgers 2000). Governments have been engaged in a process of 'competitive deregulation' (Bendell 2000: 20), seen as a 'massive transfer of assets to the private sector . . . bringing business to the heart of economic development and decision-making in almost every country . . . [and] playing an increasingly pivotal role in determining economic, social and environmental progress around the world' (Nelson 1998: 22). Traditional perceptions of political and economic institutions as separate entities have disintegrated, with greater recognition of corporate power as a shaper of political realities (Bendell 2000). As Body Shop[1] founder Anita Roddick (2000: 8) asserted:

> . . . business is the dominant socio-political force on the planet, so we can no longer pretend that the economic bottom line is the only thing that matters. If we did, then the organisations most able to make a difference to life would be shirking their responsibilities.

NGOs also have gained in this redistribution of political power as 'politicians increasingly accede their traditional power either to the corporates themselves . . . and/or to an increasing range of NGOs' (Rodgers 2000: 42). A shift in the balance of power between the state, the market and civil society has been recognised, with simpler and more traditional power hierarchies being replaced by a 'multi-relational' balance of power 'where citizens and companies are playing an active role in shaping socio-economic change and addressing problems that were previously the sole responsibility of government' (Nelson and Zadek 2000: 7). The link to NGO engagement follows, as

> it is arguably as a result of this renegotiation of the relationship between state and market that NGOs are increasingly targeting their advocacy at multinational companies – because governments increasingly seem unwilling or unable to regulate the conduct of transnational corporations (TNCs) themselves.
>
> (Newell 2000: 32)

A vital role has thus emerged for civil society as a whole to act as a countervailing force to corporate power and excess, imposing regulation through social pressure rather than legislation (Schwartz and Gibb 1999).

NGOs have developed their tactics to the point where the quality of their research, grassroots involvement, knowledge, and capacity for spectacular political action represent a counterpoint to the access powerful corporate leaders and lobbyists have to the inner sanctums of power at major international meetings (Picciotto and Mayne 1999). NGOs have used campaigns effectively to become among the most powerful stakeholder groups or external influences on companies (Rodgers 2000). In some contexts, civil society groups now effectively 'grant' a social licence to operate alongside normal regulatory mechanisms (Warhurst 2001: 72). Where state regulation has been reduced,

Figure 6.1 'Civil regulation' through direct action: campaigning against Starbucks

NGOs have developed international behavioural norms which resemble informal regulations and from which companies everywhere find it increasingly difficult to escape (Newell 2000). Social norms increasingly outstrip legal requirements imposed on firms, and thus NGOs apply new types of power to constrain corporations in the form of 'information and images to help expose, cajole, educate and persuade the corporate sector. There is a less coercive power aimed at changing consciousness and creating mechanisms of accountability' (Newell 2000: 38).

These new forms of power, defined as 'civil regulation' (Bendell 2000: 246–8), are enforced through incentives for companies to gain social, intellectual and reputational capital, and through the risk of boycotts, direct action and ethical disinvestment, leading to dropping sales or share prices, if civil society norms are ignored. NGOs are seen to have moved into the 'trust vacuum' left by companies, greatly increasing their influence (Burbury 2000). This 'complex multilateralism', a form of global governance, can be seen as a form of cooperative international organisation somewhere between state based and society-based models, necessitating the reconstitution of civil societies on a global scale (O'Brien 1999: 261). The regulatory role of NGOs has been recognised within the business world. Thus, a senior executive of the transnational oil company BP accepted that a new form of governance emerging from 'codes and accountability systems, which are becoming a kind of "soft law" . . . is being developed and enforced by increasingly sophisticated civil society activism' (Bendell 2000: 246).

The governance role of NGOs is varied. For Mori (1999), writing about environmental NGOs, this had three functions: the *presentation of global norms* or moral interests, particularly by transnational NGOs unconstrained by national or economic interests; *direct input into inter-governmental negotiations* through direct participation as well as lobbying and public campaigning; and *making new kinds of linkages*, global to local, and across sectors from the scientific to the political and from the environmental to the social. Others saw a reciprocal process in motion, with the corporate community playing a more active, although quieter, role in the development of policy and practice within civil society organisations. This was ascribed partly to 'greater interaction, trust and intimacy' and in part to 'shifts in conditionality of funding, particularly in public contracts for infrastructure development, service delivery and management' (Nelson and Zadek 2000: 55).

The move towards greater NGO–corporate engagement has been part of a more general growth in the concept of partnership. Also referred to as 'intersectoral collaboration' and 'multi-stakeholder partnership', the concept has grown in the aftermath of privatisation and deregulation (Nelson 1996: 49–54). This has been encouraged by governments, in relation to aspects of welfare, as 'public–private partnership'. A great deal of discussion about partnerships for development has occurred, some of it stemming from recognition at UN conferences in Rio and Copenhagen that government, business and civil society must work together if development targets are to be met (Murphy and Coleman 2000).

Governments, markets and civil society working individually have failed to produce far-reaching or sustainable improvements to problems related to health, nutrition, housing and education, leading to them trying joint under-takings (Kalegaonkar and Brown 2000). Such partnerships can tackle complex or large-scale issues that no individual sector has both the resources and the ability to manage. At the same time, each sector has a stake in their resolution. Three common issues have encouraged intersectoral partnership in recent years: the finance industry and economic development; environmental concerns; and 'public' issues like health and education (Waddell and Brown 1997: 5). Effective partnerships can produce innovative products, delivery systems and management. One example has been micro-enterprise lending, involving the development by banks and local community organisations of new lending products and delivery vehicles targeted at non-traditional clients who typically have little access to credit. Access to capital is thereby increased for those producing goods and services locally, often as part of NGO livelihood and community development projects (Waddell and Brown 1997).

NGOs exist in an environment of increased competition in their particular marketplaces, being forced to use more and more sophisticated techniques to communicate their messages to the public, whether these are critical campaigns or appeals for donations. Governmental and institutional donors demand higher and higher standards of reporting and administration. Some agencies in the Netherlands now maintain ISO 9000 certification[2] in order to ensure that government aid funding is professionally administered (NGO interview, 2002). As NGOs grow in size, they are also institutionalising and professionalising all manner of organisational functions such as planning, financial monitoring, human resource management and marketing (Nelson 1996; Elkington and Fennell 2000). Nonetheless, NGOs recognise that the corporate sector is generally more advanced than the NGO sector or, indeed, the governmental sector. Cooperation with corporations, particularly joint project work, has the benefit of transferring direct experience of professional financial, administrative and technical approaches to the NGO, as well as efficient methods of producing short-term outcomes (Waddell 1997; Heap 2000). Cooperation with companies can lead to better access to and credibility with governments, international bodies, other parts of the corporate world, and the supply chains of that company in order to increase influence and opportunities for reform (Waddell 1997), while greater access to the company itself enables NGOs to achieve greater leverage on other issues (Elkington and Fennell 2000).

The oldest and most fundamental relationship between NGOs and corporations is still an important one: companies are a potentially significant source of revenue for NGOs, but many NGOs acknowledge that the full potential of corporate philanthropy has yet to be realised (Sogge and Zadek 1996). When official governmental aid to NGOs is in decline, income from corporate donations is an increasingly important motivation for better NGO relations with the sector (Waddell 1997). One leading NGO's basis for relations with the corporate sector was built upon a strategic plan priority of increasing income by 50 per cent in five years (NGO interview, May 2002).

Engagement with NGOs: what's in it for the companies?

> If the institutions of democracy and capitalism are to work properly, they must coexist with certain pre-modern cultural habits that ensure their proper functioning. Law, contract, and economic rationality provide a necessary but not sufficient basis for both the stability and prosperity of post-industrial societies; they must as well be leavened with reciprocity, moral obligation, duty toward community, and trust, which are based in habit rather than rational calculation. The latter are not anachronisms in a modern society but rather the sine qua non of the latter's success. There is no trade-off, in other words, between community and efficiency; those who pay attention to community may indeed become the most efficient of all.
>
> (Fukuyama 1995: 11, 32)

The concentration of power and wealth represented by globalisation is being confronted by a growing opposition movement, which is also global in scope. Some of the most high-profile protests are now focused on multilateral bodies such as the World Trade Organization, the World Bank, the Organisation for Economic Co-operation and Development, the European Union and the

G8 meetings. Activists have frequently asserted that these organisations are controlled by big business and use their global authority primarily in the interest of big business (Juniper 1999; Newell 2000). The collapse of Enron and WorldCom due to financial dishonesty, and shareholder revolts in Britain and elsewhere against massive executive pay packages, focused attention on questions of the ethics and social responsibility of giant international corporations. More radical critics view transnational corporations as an unmitigated anti-social force responsible for pollution and poverty. A number of popular campaigns target specific companies directly, such as those involved in the production of genetically modified seeds or branded sports goods produced by poorly paid workers (Bendell 2000; Klein 2000; CDCAC 2002).

Activism in pursuit of improved corporate performance was historically centred on the producer politics of workers organised into trade unions. This is now accompanied by consumers uniting to demand change as well, often using NGOs as their organisational tools (Bendell 2000). Alongside growth in critical activism, the concept of stakeholder power has developed, as has a broadening of the definition of the responsibility of companies to their stakeholders. Primary stakeholders are no longer only the company's shareholders, but also customers, employees and suppliers. A secondary group of stakeholders, also influenced by or influencing the company's business, is being accepted: communities, governments, NGOs, the general public and even future generations (Nelson 1996; Rodgers 2000). Shareholders themselves have changed. In large public companies, many shareholders are pension fund and mutual fund managers – guided in turn by employees, trade unions, and consumers whose concerns extend into the social and environmental sphere. Shareholders can now be stakeholders, and different types of stakeholders can be shareholders. They don't want to hear that their investment is harming the environment or exploiting children in the Third World. Shareholders now include people who would willingly join consumer boycotts (Simpson 2002). Some institutional investors are pressuring corporations to take a more ethical approach to their work, a move from companies driven by shareholders to companies driven by stakeholders (Nelson 1996).

NGOs have also realised that institutional investors are an important pressure point in corporate lobbying. Activists increasingly confront investment banks and major investors directly regarding the impact of those companies in which they invest (Bendell and Lake 2000). Research from the US ethical investment industry body, the Social Investment Forum, found that one in every eight dollars under professional management in the US is now part of a socially responsible portfolio, a sum equal to US$2.2 trillion (Monaghan 2002: 142). In the interests of avoiding financial risk, shareholders and lenders may also pressure companies to display social and environmental responsibility. Share prices in developing country markets move up following positive press reports of companies' environmental performance and down in response to news of pollution incidents or fines (Dasgupta *et al.* 1998). Thus, social responsibility can be seen in terms of 'corporate governance' and the extension of this from simply good working relations between the managers, directors and shareholders to a broader network of stakeholders (Nelson 1996: 43).

In part owing to the wake-up call of critics both on the streets and in the shareholders' meetings and in part through self-awareness of global trends, a number of more far-sighted corporate leaders are now aware that their growth in power and prominence increases their need to demonstrate social and environmental responsibility. Conscious of the validity of some of the criticisms levelled against them, they seek to change their companies' and sometimes their sector's ways of working, seeing both short-term and long-term enlightened self-interest in an ethical stance for the company and the more general wellbeing of society. Companies with the capacity to predict tightening regulations on social or environmental conduct can take early action and gain 'first-mover advantage' over those companies forced to take remedial action at the last minute (Hutton and Cowe 2002: 89).

As the leading Philippine businessman Sixto Roxas III acknowledged: 'to the extent that the businessman's [*sic*] economic activities generate an imbalance in society and create social tensions, he must undertake social development programs which respond to these problems' (Waddell 1997: 12–13). Capitalism itself may be unsustainable unless companies play their part in addressing the widening gap between rich and poor people and both rich and poor nations, for 'Those who feel ignored or un-cared for are hardly likely to fulfil their side of the sustainability bargain' (Elkington 1997: 346).

Risk avoidance has been a driving factor behind increased transparency on environmental and social performance. The damage to business in being seen to behave badly is far more significant than the gains from doing good; thus 'eliminating the negative currently takes precedence over accentuating the positive' (Hilton and Gibbons 2002: 102). Three types of risk result from poor environmental and social performance: reputational, operational and legal (Bendell and Lake 2000). A study by the Social and Ethical Risk Group, formed by a group of British companies, concluded: 'Risks occur when a gap opens up between company behaviour and society's expectations' and 'from practices that accidentally or deliberately harm the health or well-being of people or society, such as labour issues, bias or social exclusion' (Sutcliffe 2001: 14). NGOs can help companies reduce potential risk by integrating business and community goals and fostering inclusive and transparent consultation, dialogue and negotiation. They can assist in creating, implementing and monitoring popularly supported standards and codes and help with community education and training. With contacts, networks and alliances outside the corporate structure, NGOs can articulate the concerns of more indirect stakeholders or provide early warning of potential problems with corporate activities (Waddell 2000). Identifying a damaging business venture at the outset through assessment against environmental and social criteria is likely to be less costly in both economic and reputational terms than being forced to abandon a project after a confrontational public campaign by critics. Early participation of local NGOs and communities can obviously help in avoiding such costly mistakes and, more broadly, enable companies to become more socially responsible (Heap 2000).

The brand has become the principal asset of more and more companies as a greater proportion of costs go towards marketing and support functions and the production costs of any product become a smaller proportion of the final price (Korten 1995). Product quality of different companies is increasingly indistinguishable, whilst the image suggested by the brand has become the major distinction. Major companies such as Nike don't actually manufacture their products; these are subcontracted to faceless manufacturing companies in the developing world. Nike's principal work is to create and market its image, so its clothes and shoes can be sold at a premium (Klein 2000).

As the brand becomes a company's principal asset, threats to that brand caused by poor records on social, environmental or labour standards become a potent new kind of business risk, representing a threat to 'reputational capital' (Bendell 2000: 23). Campaigners against corporate abuses have not been slow in realising the new vulnerability companies face with regard to their brand. Many campaigns specifically directed at a brand have proved effective at forcing corporate targets to alter their policies and practices. Thus, 'Brand image, the source of so much corporate wealth, is also, it turns out, the corporate Achilles' heel' (Klein 2000: 343; Linton 2005).

A brand identity can be established through shrewd marketing and innovative advertising. Good reputation with stakeholders, however, is built on the application of values into everyday business practices and is therefore much more difficult to establish, and more difficult to restore when damaged (Peters 1998). Around one-third of shareholder value in many sectors of industry is attributable to company reputation (Kearney 1999; Bendell and Lake 2000). Some 37 per cent of UK consumers were influenced by ethical concerns such as fair trade, the environment and world poverty; 24 per cent avoided services or products from companies thought to have a bad environmental record; and 52 per cent put off making a purchase because of animal welfare issues (Thomas and Eyres 1998: 11).

Reputation also affects the internal dynamics of a company. Staff motivation and morale, and the ability to recruit and retain more talented staff, increase in companies with favourable public reputations and strong social programmes. People working longer hours are left with less time for citizenship and community activities (Draper 2002). Decision makers are becoming unwilling to compartmentalise their lives, their behaviour or their values between work and home, wishing instead to feel satisfied and fulfilled by all aspects of their work (Hilton and Gibbons 2002). A key part of the 'social capital' of a company includes staff who are motivated, knowledgeable and skilled, critical in the modern corporate world (Bendell 2000).

Major international companies dwarf not only all NGOs but also the majority of the world's nation states in terms of economic power. Yet the internet's capacity to democratise both the receipt and the dissemination of information has helped change the balance of power between companies, NGOs and consumers. The revolution in information technology means the general public now hear of corporate misconduct more quickly and directly than ever before, and people affected by corporate activity can communicate more easily and effectively with that company's stakeholders. Communities displaced by mining companies, or garment workers forced to work overtime in dangerous conditions, for example, can explain their grievances to developed-world customers or fellow workers at the head office quickly and convincingly, sometimes with internet facilities provided with the assistance of NGOs (Bray 2000; Klein 2000). Therefore, 'In an interconnected world, companies are coming to realise that there is no hiding place for poor performance on environmental and social issues' (Bray 2000). Media surveys show coverage of NGOs quadrupled between 1996 and 2000 and, at the Seattle WTO meeting, reported quotes from NGOs were more than double those from the WTO itself (Lombardo 2000).

The use of new information technology has become potent. During the McLibel trial, when McDonald's sued some activists who had published a pamphlet criticising the restaurant chain's labour, social and environmental practices, McDonald's critics launched the McSpotlight website to present the huge amount of information they were receiving from other activists and McDonald's employees around the world. The site also contained the text of the offending pamphlet, and the complete 20,000-page transcript of the trial, and became one of the most popular destinations on the web, accessed approximately 65 million times (Klein 2000). In January 1999, a group of environmental activists occupied part of Shell's UK headquarters. A digital camera and a computer linked to a mobile telephone allowed them to broadcast their sit-in on the web and to email journalists, even after Shell turned off the electricity and cut the telephones (Vidal 1999).

Companies which don't sell directly to the public can shrug off controversy, while 'Multinationals based in Asia fly below the radar of Western activists, and growing local protest movements have tried but failed to stir up outrage against huge Indonesian mining operations and Malaysian logging firms' (Miller 2001). However, this may not remain the situation much longer because consumer vigilance is growing and campaigners are becoming more willing to take on complex industries. Environmental groups such as the National Wildlife Federation and Friends of the Earth are targeting banks that 'fuel the chain saws' after research on financial records, and launched the 'Spank the Bank' campaign criticising Citigroup for issues which included financing the controversial Three Gorges Dam in China.

Effective links with NGOs and civil society bring direct business benefits to companies. Multi-stakeholder partnerships at every level of society help corporations identify the political, social, cultural and environmental opportunities offered by a country, all of which lead to economic strength and potential (Nelson 1996). By working with NGOs, companies discover important new perspectives about the issues of concern not only to local communities but also to the media and regulators (Plante and Bendell 2000). NGO skills useful to business include consensus building, culturally sensitive dispute resolution, creating and using voluntary participation, ways of working with specific racial, ethnic or low-income communities, aspects of stakeholder management and communication (Waddell 1997; Heap 2000).

Educated and media-attentive 'thought leaders' in the US, Europe and Australia were 'two to three times more likely to trust an NGO to do what was right than a large company, government or the media'. Between 1955 and 1995 there was a 50 per cent reduction in trust and confidence in big business to do what is right for employees, consumers and people in general, while most respondents to a survey believed NGOs were motivated by morals rather than profit. More than 75 per cent of respondents believed corporations should speak out on issues such as the environment, human rights and health care (Lombardo 2000: 4, 22–9). The Australian public viewed NGOs as 13 times more credible than companies on environmental issues and more than 20 times more credible on human rights (Monaghan 2002: 140). Relationships with NGOs can therefore enable corporations to re-establish public trust.

Companies working with NGOs can also benefit through the identification of new markets and in the development of products suitable for such markets, for 'Community knowledge is critical in creating new products for particular demographic and psychographic profiles' (Waddell 2000: 197). NGOs provide such knowledge, as they are closer on a day-to-day basis to grassroots communities which may consist of ethnic, racial or low-income groups with potential as new corporate customers (Waddell 1997; Elkington and Fennell 2000). NGOs may, in some situations, be able to use their programmes and their networks to aggregate small markets into a profitable size and leverage both tax breaks and volunteer resources. The presence of a respected local NGO in a project may also reduce the potential for vandalism, crime and corruption (Waddell 2000).

Changing concepts of corporate responsibility

Three areas merit analysis for understanding better the contributions of business to sustainable development (Warhurst 2001: 61). These are *product use* in society and the contribution of industrial products to improved health, wellbeing and quality of life; *business practice*, involving the way in which the business is run (corporate governance and in particular the extent of social (including environmental) responsibility integrated within corporate strategy; and finally *equity* and the intra- and inter-generational distribution of the benefits of industrial production across different societies, especially within host communities. While the first two criteria may be within the power of the individual company to control, equity is an element of a broader debate in political economy, beyond the control of an individual company. A more practical categorisation of the role companies can play in the development process identifies the *core business activities* of a company, its *social investment and philanthropic activity*, and the company's government relations and *engagement in public policy dialogue* (Nelson 1996: 58–75).

Until recently, companies usually demonstrated good citizenship through corporate philanthropy, customarily handled by staff attached to public relations or community affairs departments, or the corporate foundation. Philanthropy was a marginal activity, with separate budgetary allocation, and largely unrelated to the principal activity of the organisation. Material on such work either appeared in staff newsletters or was pushed out by publicity departments in the hope of favourable mention of the company name in the media (Nelson 1996; Hilton and Gibbons 2002). In recent years, a number of major companies have changed their perspective on charity. Social involvement has become part of their core strategy, infusing the entire corporation and forming part of corporate strategic management, philosophy and identity. Some mining companies, for example, have launched educational, agricultural, health and environmental programmes in the communities affected by their mining. Accompanying this change in external relationships have been efforts by companies to improve social and environmental conduct related to their core business activities. Such programmes were accompanied by examination of the core philosophy and mission of the company and often accomplished through a fundamental re-orientation or re-branding. This sometimes followed a period of stinging criticism. The re-branding of BP from the oil company British Petroleum to an energy company, Beyond Petroleum, with a new logo in the form of a flower/sun, provides an example (BP 2001), though this might also be seen as simple 'greenwashing'.

Most large companies have instituted departments or senior management posts concerned with social and environmental responsibility, and introduced social reports, which indicate progress against the 'triple bottom line', adding social and environmental performance to the traditional bottom line of economic performance. Companies also voluntarily introduced codes of conduct detailing the labour and environmental standards they would apply in manufacturing and subcontracted supply of their products in developing countries (BP 2001; Shell International 2001). In some cases, companies have modified their commercial activities to meet development needs, while still aiming to profit from the results. A number of banks, for example, offered discounted banking services as well as financial training to micro-enterprises in developing countries (Prahalad and Hart 2002). Some companies offered their core competencies in organisational and management skills to social projects, rather than simply donating cash. Information technology companies have supplied free technical advice, equipment and staff to disaster and refugee situations. Cable and Wireless worked with CARE UK, for example, to produce emergency communications kits (Nelson 1996). The Swedish company Ericsson supplied volunteer technical staff to help repair basic communications systems as well as providing training and donating equipment in emergency situations such as the earthquakes in Turkey and Gujarat, India, and during the refugee crisis in Kosovo. More recently, Nokia provided mobile phones to rescue teams following the South-East Asia tsunami disaster and worked to expand and restore services, as well as providing a cash donation to the Red Cross and establishing a €2,500,000 reconstruction fund (Nokia 2006). Some pharmaceutical companies earmarked manufacturing inventories for donations, and work with relief agencies on contingency planning for disasters. Some corporations producing large amounts of waste donated funds to NGOs for projects which offset the impact of their pollution by reducing the equivalent amounts of carbon emissions. Certain car manufacturers, electricity generation companies and airlines, for example, have funded reforestation projects and energy renewal and energy efficiency projects to offset their carbon emissions (Hilton and Gibbons 2002).

Through changes in core business orientation, business can discover lucrative new markets while contributing to poverty alleviation and equitable development. Five fundamental assumptions that corporations hold can be contested: the poor are not the corporate target; the poor cannot afford products; only developed markets will pay for new technology; the bottom of the market is not important for long-term corporate interests; and the intellectual excitement is in the developed markets (Waddell 2000; Prahalad and Hart 2002). Multinational corporations can put into place conditions that enable the poorest parts of society in the developing world to improve their livelihoods. These corporations have the resources, managerial skills and knowledge necessary to build commercial infrastructure and are best positioned to unite NGOs, communities, local governments and local entrepreneurs to meet the needs of the poorest sector of the market (Prahalad and Hart 2002).

The achievement of corporate responsibility is still in its early stages, and it remains hard to gauge the extent of the growth of this aspect within the corporate world. By the mid-1990s, more than 100 of Standard & Poor's top 500 companies were producing environmental reports for shareholders, but in 1998 less than 1 per cent of 34,000 transnational companies were active in the debate on corporate social responsibility (Hutchinson 2000). Already, according to some analysts, corporate social responsibility is an ageing concept, or at least an ageing term, to do with risk reduction and philanthropy. Cutting-edge companies have been moving towards 'corporate social leadership' or 'corporate social investment' in which companies 'find a dual purpose, social as well as commercial, for every component of the corporate anatomy' (Hilton 2002). For corporate critics, on the other hand, corporate social responsibility is a voluntary smokescreen, a new kind of 'greenwash', attempting to ward off mandatory standards. They call instead for 'corporate social accountability' centred around national and international regulation.

Forms of engagement: NGO–business interactions

The earliest interaction between NGOs and companies took the form of requests for donations and material support. This uncritical relationship corresponded to an early stage in the development of NGOs, which then either primarily were involved in famine and disaster relief or viewed work on poverty and underdevelopment in terms of the provision of welfare (Smillie 1995; Sogge 1996b). As development NGOs became more aware of the underlying causes

Figure 6.2 NGOs have long protested at the marketing of infant formula in developing nations

of poverty, they began programmes to 'help people help themselves', to build local capacity and to examine the social and political barriers to equal opportunity for the poor and dispossessed. NGOs also became more vocal in advocacy about the causes of poverty and vulnerability, which inevitably led them towards critical campaigns against company policies and practices. Early examples of these included the campaigns against the promotion of powdered baby milk in developing countries and criticisms of companies investing in apartheid South Africa (Smillie 1995; Fowler and Biekart 1996; Sogge 1996c).

While fundraising from companies and critical advocacy both remain a significant part of development NGOs' work, the newest area for interaction between companies and NGOs is that of engagement in order either to jointly pursue social programmes or to improve the core business practice of the company. A transition in NGO relationships with corporates has occurred 'from one of confrontation to one of consultation and even association' (Rodgers 2000: 41).

NGO engagement with the corporate sector can be classified into broad categories: donors and marketing relationships; critical advocacy; lobbying and campaigning; programme work on codes of conduct, industry standards, fair trade and social auditing; and cooperation in development and relief work (Elkington 1997). For NGOs, the mix of relationships can be complex. Receipt of donated money from, or project cooperation with, companies clearly sits uneasily with public criticism of their conduct, let alone campaigns questioning the underlying ethos of corporate activity such as foreign investment, market liberalisation and international financial systems. Only one example of an NGO simultaneously criticising a company on one issue while cooperating with it on another could be found. The environmental group Greenpeace worked with Monsanto on the production of non-PVC credit cards while campaigning against its policy on genetically modified foods (Heap 2000).

NGO policy towards the corporate sector divides them, according to Elkington, into *'polarisers'*, which are basically critical and confrontational, and *'integrators'*, which attempt productive relationships with corporations. In addition, NGOs are either *'discriminators'*, which pick on specific companies, or *'non-discriminators'*, which focus on an entire business sector (Elkington 1997: 228–30). While such dichotomies provide a useful analytical perspective, this takes little account of the many NGOs with a variety of approaches towards companies. NGOs might conduct a critical campaign against an entire sector, such as weapons manufacturers, while working closely with an individual company in a different industry on specific improvements to its labour or environmental standards. On other issues, such as infant formula or pharmaceutical provision, they may decide that the best strategy is to single out the worst offender for a single-company campaign. Thus, NGOs willing to work together with companies on development projects could therefore be distinguished from those which choose to work with companies to change their core business practices.

Bendell's four-category typology of NGO–business relations divides NGO activity into that dependent on raising revenue within the market and that working outside the market, and also separates work between that of a confrontational style and that of a collaborative approach (2000: 242–5). NGOs working in the market in a collaborative style are *'facilitating change'*, engaged in such activities as consultancy and monitoring of standards. Those working in the market in a confrontational style are *'producing change'*, engaging in activities like fair trade and other alternative economic schemes. NGOs that are collaborative, yet outside the market, are *'promoting change'* by working on such things as codes of conduct and standards. Finally, NGOs working outside the market in a confrontational style are *'forcing change'*. This category includes the campaigners, demonstrators and boycotters critical of corporate activity.

A number of corporations are working in partnership with international NGOs on long-term development, social and conservation projects or the creation and implementation of standards or codes of conduct. Each side invests knowledge, staff and finances to achieve agreed goals. Such 'strategic partnerships' involve the core business or programme activities of both partners, and differ from past forms of relationships, being neither philanthropic relationships in which businesses simply donate funds to NGOs nor adversarial relationships based on NGO protests against corporate behaviour (Ashman 2000). In programme cooperation with corporations, NGOs are often seen as intermediaries, building bridges between corporations and a variety of stakeholders, including affected communities, potential customers, and others who wish to voice their opinions to the company (Waddell 2000). Clear roles can be identified for NGOs in fostering intersectoral cooperation between Southern NGOs, business and government, including strengthening local capacity, training on the practicalities of intersectoral cooperation, lending support to pilot projects and contributing to risk reduction (Kalegaonkar and Brown 2000).

Much collaborative work involving NGOs has emerged from cooperation on the creation of standards and codes of conduct, for either a single company or a whole industry; for example, the Forest Stewardship Council covers producers and retailers of wood and other forestry products. In designing, developing and implementing codes and standards, companies have sought to involve NGOs and sometimes their critics. In the design and implementation of codes of conduct, there are a number of possible roles for NGOs, including: watchdog; auditor; mediator (between workers or communities and auditors); clearing house for reporting problems; remediation agent (supporting those negatively affected by codes of conduct); and trainer of auditors on local conditions and laws (Tepper Marlin 1998; Heap 2000). Britain's Ethical Trading Initiative brings together the government, NGOs, companies and trade unions to develop joint labour and social codes in global supply chains, as does the Fair Labor Agreement in the USA and the Canadian Taskforce on labour standards in the textiles and footwear sector. As WTO rules now prohibit discrimination against imports on the basis of the conditions of their production, such initiatives become a way of raising standards through consumer, NGO and company peer pressure (Kearney 1999; Williams 2005).

Among the most fundamental criticisms of voluntary codes and standards is that they reduce the legitimate role of governments in regulating production (see Box 6.1). Responsibility is handed to unelected and unaccountable business and civil society organisations. A number of studies suggest companies create codes of conduct to avoid stricter government regulation (Heap 2000; Klein 2000). As veteran international trade unionist Neil Kearney argued: 'Corporate codes of conduct, in effect have privatised the implementation of national labour legislation and the application of international labour standards' (1999: 208). The harshest critics view corporate codes of conduct as no more than the products of corporate public relations departments, emerging as panic measures after embarrassing media exposés, designed to subdue critics. Campaigner Naomi Klein was dismissive: '[W]hat we have with the proliferation of voluntary codes of conduct and ethical business initiatives is a haphazard and piecemeal mess of crisis management' (2000: 430). In response to this long list of criticisms, proponents of codes of conduct argue that their creation need not obviate campaigns for stronger government regulation. Companies with strong codes, in fact, have reason to become allies with NGOs in this work to avoid rogue companies and 'free riders' undercutting their more responsible business practices. Thus, in pursuit of a level playing field, companies with better environmental and social records could become allies with NGOs in pushing governments for a levelling up rather than a levelling down of regulatory standards. As arguments on the merits of codes of conduct continue, the opinion of several 'corporate activists' in NGOs, including those active in the promotion of voluntary corporate and industry-wide codes of conduct in recent years, has now come full circle. There is a growing belief that NGOs should apply more energy to strengthening national legislation and pushing for effective implementation of existing legislation. What has changed is that NGOs may find themselves able to join with certain companies in their approaches to government (interview with Oxfam GB policy maker, 2002).

For many NGOs, the principal relationship they have with corporations remains that of a recipient of major corporate donations. These donations may take the form of not simply cash but also goods of use to the agency, such as computer equipment or office furniture, or of value to the beneficiaries the agency is seeking to help, such as blankets, medicines, school books or beds. Companies also donate services, such as management training, technical advice, marketing and communications services or professional auditing and accounting (Heap 1998). Over-dependence on official donations has raised questions about the independence and autonomy of development NGOs. Several NGOs see income from the corporate sector as the least exploited area of possible income expansion (compared to public fundraising) and any increase in income from companies represents a move away from over-dependence on governments (Smillie 1995; Saxby 1996). In addition to philanthropic donations, NGOs have developed a number of marketing relationships through which they gain revenue

Box 6.1: **Criticisms of NGO involvement with codes of conduct**

Criticisms of NGO involvement with codes of conduct raise a large number of questions, including:

- NGOs lack the capacity, authority and legitimacy to design or monitor such standards (Heap 2000; Murphy and Bendell 1997).
- Work on codes can draw NGOs into complex organisational relationships with business, draining them of vitality and flexibility (Nelson and Zadek 2000).
- It is more important for NGOs to use resources to retain pressure on corporations through adversarial lobbying and public campaigns (Fowler and Heap 2000).
- By helping companies design, implement, monitor and evaluate codes of conduct, NGOs are playing the role of legislator, police force, judge and jury, which is unsatisfactory.
- Some NGOs will take payment from corporations for work on codes of conduct. This presents the danger of co-option, or the emergence of NGOs set up simply to bid for such work. Critical objectivity is endangered by such a financial relationship (Heap 2000).
- Problems with devising effective systems for monitoring and compliance seem to be a principal weakness in codes of conduct in general. Where monitoring plans exist, NGOs and unions are often excluded (Ferguson 1998; Picciotto and Mayne 1999).
- The lack of worker participation in the creation and monitoring of codes of conduct is seen by labour organisations as their biggest weakness (Heap 2000; Asia Monitor Resource Centre 2001; O'Rourke 2001).
- Codes of conduct can force a centralisation of production in larger plants with the resources to operate to the required standards, removing the livelihoods of homeworkers and pieceworkers, who are usually women and sometimes children (Newbold 2002).
- Vulnerable groups such as women and immigrants are seldom accounted for in the labour provisions of codes of conduct. Where codes enforcing equal standards are applied in a 'one size fits all' manner, there is a danger that women would be the first to lose their jobs (Ferguson 1998).
- Codes can impose a Northern consumer agenda. For example, they can attach more importance to issues of child labour than to the more complex and fundamental issue of the right to organise trade unions or the right to a livelihood (Ferguson 1998). For obvious reasons, they also tend to concentrate on export industries only (Heap 2000).

- Large transnational companies adopting codes of conduct can oblige subcontractors to apply the standards and criteria. This can push the costs of compliance on such items as environmental standards on to contractors, who also remain under pressure to produce goods at the cheapest prices. Subcontracted suppliers will naturally look for savings in other areas, often in terms of wages paid to their workers (Kearney 1999; Heap 2000).
- As codes and standards are often multi-party agreements, there is a danger that the ultimate agreement will consist of weak provisions representing the lowest common denominator (Nelson and Zadek 2000).

in exchange for providing direct publicity for the corporation. This can take the form of well-publicised sponsorship of NGO events, or matched giving programme endorsement of a particular product in exchange for a donation (Heap 1998). While this relationship is intended as a 'win–win' form of cooperation, there are certain risks for NGOs, in that they may be seen to be endorsing whatever product is offered associated with their name (Smillie 1995; Sogge and Zadek 1996).

The most prominent type of NGO–corporate interaction is that of critical advocacy. This often involves high-profile public and media campaigns designed to pressure companies by changing public perceptions of them, and often includes a specific set of demands. The majority of NGOs are investing increased resources in this area, in the media and communications staff necessary to propagate an effective message, and through research and policy teams working on the background analysis necessary to make effective and supportable cases (Anderson 2000). Thus, while many NGOs do not have the strength or resources to impose criteria on corporations

> . . . they are becoming adept at mobilising other more powerful stakeholder groups to take up their position: that is, they are operating as stakeholder catalysts. Consumer boycotts, media pressure and moral outrage are but a few techniques being employed to incite stakeholders to 'take up arms' against specific corporate activities.
>
> (Rodgers 2000: 45)

Environmental and labour groups are stronger in their advocacy of international standards and enforcement mechanisms, whereas NGOs are more ambivalent about such instruments as social clauses in international economic agreements (for example in WTO rules), fearing these will be used by wealthier states to economically disadvantage developing countries (O'Brien 1999). Among development agencies, approaches differ. Some believe that NGOs may indeed increase cooperation with corporations in future, but that critical pressure should also be maintained. While collaborative NGOs should not engage in initiatives that whitewash or distract from critical NGO campaigns, oppositional NGOs should avoid setting unrealistic standards for cooperation (Heap 2000).

Policy challenges and NGO–corporate interaction

Three potential costs for NGOs are associated with collaboration: the time and energy invested in learning to relate to the corporate partner; adapting the organisation to meet demands of collaboration such as tangible and timely results; and reduced effectiveness resulting from the lack of influence in the partnership. The costs of adapting to collaboration are usually borne by the NGO rather than the corporate partner (Ashman 2000).

Cultural differences feature in the negotiation process between NGOs and companies, with companies more prepared to bargain, because they are used to negotiating across mixed interests. Civil society organisations, with their values- or rights-based analysis, frame differences in ideological terms, leading to polarised position taking based on rights or power, rather than compromise and bargaining based on critical cooperation. NGOs are also under pressure from more complex and contradictory stakeholder pressures than their business counterparts. When business and NGOs form a partnership, therefore, they may create more complex sets of external stakeholders, which subject them to unforeseen demands and pressures (Covey and Brown 2001).

Among the most fundamental issues facing NGOs with regard to their policies on the corporate sector is the classic dilemma of choosing between pragmatism – the making of realistic short-term gains – and adherence to the dearly held belief that sustainable change needs to be more radical and far-reaching. In practical terms, this becomes a choice between prioritising work with companies on their programmes of internal change and social responsibility and emphasising critical advocacy pushing for deeper changes in national and international regulatory frameworks and the international economic architecture.

Many remain sceptical that corporations, in their present form and within the present economic system, can truly develop into sustainable, socially positive organisations. They therefore advocate that NGO energy be concentrated firmly on campaigns for macro-economic changes targeting state and multilateral bodies and that continued critical activism against the corporate sector consist primarily of making the case for regulation. They would not support an increase in programmatic relations between NGOs and corporations, doubting that NGOs can do any more than apply sticking plasters to an inherently unjust economic system with corporations at its heart. These critics of cooperation suggest that working with TNCs consolidates corporate power, increases their standing in society, endorses their activities, and strengthens their defences against attack from more critical NGOs (Heap 2000). Instead, TNCs should be ceding power, as they are the underlying cause of social and environmental problems (Nelson 1996; Heap 2000; Covey and Brown 2001).

Companies' efforts towards more social responsibility can make them vulnerable to mergers and takeovers by larger and more aggressive companies, which might result in the erasure of any social responsibility programmes (Elkington and Fennell 1998). Barriers to responsible corporate behaviour are inherent in the economic structures, despite perceptions of 'first-mover advantages':

> A rogue financial system is actively cannibalising the productive corporate sector. In the name of economic efficiency, it is rendering responsible corporate management ever more difficult. Those who call on corporate managers to exercise greater social responsibility miss the basic point. Corporate mangers live and work in a system that is virtually feeding on the socially responsible.
>
> [...]
>
> We must not kid ourselves, social responsibility is inefficient in a global free market, and the market will not long abide those who do not avail of the opportunities to shed the inefficient.
>
> (Korten 1995: 214, 237)

These articulations of anti-capitalist ideology, often incorporating ideas from feminism and environmentalism as well as socialism, see economic growth as a false god in the quest for a just and sustainable society. In an economic system dominated by market forces, growth and competition are inimical to the creation of a system of equity and justice.

Significantly, however, both reformers and revolutionaries see a vital role in the change process for civil society and NGOs. In contrast to the faith that socialists and communists of the past placed in highly structured and centralised party and party-state structures, contemporary radicals believe that change must come from people more organically organised and networked around a diversity of causes, and demanding a stronger voice and greater participation in decisions that affect their lives and an end to the remote centralisation of power (Korten 1995). This alternative is already emerging: 'Ethical shareholders, culture jammers, street reclaimers, McUnion organizers, human rights hacktivists, school-logo fighters and Internet corporate watchdogs are at the early stages of demanding a citizen-centred alternative to the international rule of the brands' (Klein 2000: 446). While such a movement can be both high-tech and grassroots, both focused and fragmented, it is global and as capable of coordinated action as the multinational corporations it 'seeks to subvert' (ibid.: 446).

A second major challenge for NGO policy making on engagement with the corporate sector concerns the capacity of NGOs to rise to all the challenges and opportunities presented by closer working relationships with companies. The number of companies seeking stakeholder dialogue with NGOs has grown considerably, while NGO resources have remained constant: 'As a result, the individuals within NGOs that are the most amenable to dialogue with companies are increasingly besieged with invitations' (Elkington and Fennell 2000: 153).

A critical policy area in NGO–corporate relations concerns the measurement of the impact of this engagement, particularly if an NGO with limited capacity is attempting to measure the relative impact of its work related to government regulation, poverty alleviation and engagement with the corporate sector. The development of indicators of relevance to corporate stakeholders is vital in this process (Bendell and Lake 2000). Thus, 'Measuring the impact of, and effectively communicating about, the partnership is therefore not an after-thought, but an essential element of an effective pathway for any successful partnership' (Nelson and Zadek 2000: 45). Unfortunately current measurement of and communication about the impacts of partnerships are often inadequate, in terms of both quality and timing.

In cooperative programmes, companies, for their part, have concerns about confidentiality of information shared with NGOs and difficulties dealing with the tendency for NGOs to extend the agenda of concern as the relationship matures. There are further concerns over inconsistency of message and approach, not only between different NGOs but even within the same NGO and even within the same individual suffering 'ideological tension' (Crane 2000: 172; Elkington and Fennell 2000). This raises policy questions for NGOs regarding external communication with the company as well as internal communication and management support.

While those within a company responsible for community affairs or corporate social responsibility may reach out to NGOs, other staff or departments may be hostile towards or suspicious of NGOs. Similarly, a significant policy challenge for NGOs is to manage their own internal conflicts within staff, membership and donors on working with the corporate sector. Handled badly, corporate partners will experience inconsistent behaviour from different staff or departments within the same NGO, as partnerships with companies accentuate or aggravate differences between fundraisers and campaigners (Elkington and Fennell 2000; Heap 2000).

Co-option or compromise of an NGO through collaboration with a large and powerful corporation is a significant issue. Whether co-option is real or simply perceived to be so by constituencies or others in the NGO community, problems arise. While too much antagonism can reduce the effectiveness of NGO–business cooperation, when one organisation subsumes its perspectives and interests to that of another the value of diversity in cooperation is lost (Heap 2000; Kalegaonkar and Brown 2000). There is a danger that NGO–business alliances 'may lead to "NGO capture" in much the same way as powerful corporates may succeed in dominating relationships with regulators giving rise to "regulatory capture"' (Rodgers 2000: 41).

A second practical challenge is resolving the question of the motivation of the corporate partner. Acknowledging that the corporate sector is diverse and multi-faceted, some NGOs believe in the need to differentiate between those which are seeking genuine change internally and externally and companies which are seeking to give the impression of change, concern and action with the principal motivation of improving public relations or deflecting criticism.

Some argue cooperative projects should be judged by their impact alone, and the internal motivation of the company is unimportant. To others, cooperative project work should be part of a longer-term engagement providing opportunities to change the outlook of the corporation.

Policies for successful NGO–corporate interaction

To overcome the policy challenges outlined above, the literature on effective NGO–corporate relations provides frameworks to enable successful cooperation: accepting differences, delineating roles, building trust and clarifying the limits to any cooperative relationship. Similar factors are considered important to the success of NGO–business collaboration:

1 a clearly defined plan, identifying common ground, common objectives and mutual gains, but also being clear where the parties agree to differ and where limits to engagement lie (Nelson 1996; BOND 1999; Fowler and Heap 2000; Kalegaonkar and Brown 2000; Weir 2000);
2 adequate resources and clarity about the costs and benefits to each party (BOND 1999; Fowler and Heap 2000; Kalegaonkar and Brown 2000);
3 awareness of, and efforts to balance, power asymmetries in the NGO–corporate relationship, achieved through recognition of the mutual benefits of the relationship or through the mediation of third-party organisations (Kalegaonkar and Brown 2000);
4 professional staff on the part of the NGO (BOND 1999; Fowler and Heap 2000);
5 internal capacity on the part of the company (Fowler and Heap 2000);
6 agreement about levels of transparency and communications strategies (BOND 1999);
7 agreement about involvement of other stakeholders of both parties, including management of potential stakeholder unease; inclusivity in identifying stakeholders is important for both parties (Nelson 1996; BOND 1999; Heap 2000; Weir 2000; Covey and Brown 2001);
8 good preparatory research by the NGO; anecdotal evidence alone is inadequate (BOND 1999);
9 leadership from senior management, as without this there is a danger that the individual innovators of partnership between NGOs and corporations will become separated from their own organisations (UNEP 1994; Nelson 1996; Austin 2000; Nelson and Zadek 2000);
10 agreement about roles, responsibilities and decision-making processes (Nelson 1996; BOND 1999; Kalegaonkar and Brown 2000);
11 a common analysis by the NGO of the role of corporations and the role of NGOs in engaging with them (Heap 2000).

Intermediaries, either individuals or organisations, can be one of the most crucial factors in NGO–business partnerships. They build consensus between parties who may bring different motivations, needs and resources to a plan for partnership, and provide 'consultative or facilitative leadership' (Nelson 1996: 275–6). Intermediaries may also provide the necessary technical training and produce efficiencies of scale to enable successful partnerships (Waddell 1997).

A study of intersectoral cooperation in Madagascar, South Africa and the Philippines emphasised the importance of finding convenors with credibility across the sectors who may be able to mediate traditional antagonisms; it often helps when these convenors are outside the sectors or organisations concerned (Kalegaonkar and Brown 2000). In particular, 'social entrepreneurs' as facilitators of innovative processes (characterised as risk-taking, innovative, energetic and creative) are critical to making partnerships work (Nelson and Zadek 2000: 40). Even where a company desires consultation, local NGOs may lack capacity to present cases effectively, and different types of NGOs, such as those defending traditional people's rights and those defending the environment, might not always be in agreement (Ali 2000). International NGOs therefore have a specific intermediary role assisting a representative range of local NGOs in making their case and developing effective negotiating skills.

Any mutually agreed rights, norms or legal frameworks should be identified and clarified in the process of forming a partnership. However, rights-based approaches are usually only appropriate in circumstances where such rights and obligations have been agreed and defined. An inflexible rights-based approach involving positional bargaining may stand in the way of reaching agreement when both parties need to resolve a dispute over socially acceptable standards. Interest-based negotiations stand a better chance of success in meeting the needs of all parties. This approach is more capable of dealing with conflicting as well as convergent interests, has lower transaction costs, can lead to greater satisfaction with outcomes, and produces more durable solutions (Covey and Brown 2001). If problems can be framed or re-framed as issues of inter-dependence, intersectoral cooperation stands more chance of success. Other important factors include the need to invest in relationship building as well as problem solving, to foster mutual influence in decision making and to prepare well for managing conflicts (Kalegaonkar and Brown 2000).

The level of formality in any collaboration needs to be considered carefully. There are advantages in establishing organisational and legal aspects of intersectoral partnerships, from the perspective of managing resources and ensuring accountability where these are significant factors. On the other hand, over-formalisation creates additional costs and can slow down decision making and constrain innovation. Thus, 'too much structure and formality can have the effect of re-separating the various participants, reducing the partnership back to its constituent parts and as a result, losing the alchemical element that is so central to their success' (Nelson and Zadek 2000: 44).

As this overview has shown, interaction between NGOs and corporations has grown in response to increasing recognition by NGOs of the central role corporations play in shaping development paths of nations and communities. This is reinforced by recognition on the part of corporations of the risks of failing to respond to social criticism, as well as the benefits of working with NGOs to become more socially responsible. This growth in NGO–corporate relationships has resulted in an increasingly complex range and style of interactions, as outlined in this chapter, each presenting challenges for NGO policy.

7 Risks and rewards

NGOs engaging the corporate sector

John Sayer

The relationships between international development NGOs and corporations are explored further in this chapter; it addresses this subject from the perspective of the policies NGOs hold towards corporations, and the processes by which these policies have been developed. In particular, four dimensions of these relationships are examined: how much common ground NGOs share in terms of their interaction with companies; the coherence of NGO calls to companies for improved conduct and a greater contribution to social justice; the internal process of NGO policy making regarding corporate relations; and the tensions which may have emerged in the policy-making process.

To anchor a very open-ended subject about relations between two major sections of society (civil society and the corporate sector), the research on which this chapter is based focused on policy making by NGOs concerning their relations with corporations. NGO policy documents therefore formed the heart of the research, representing the finished products of a complex policy formation process. However, written policies reveal little about problems with their practical implementation, so other sources of information were sought, including third-party studies of NGO–business relations (mainly academic) and information from interviews with a range of NGO staff. The latter were the principal focus of the research. Interviews were conducted with 25 staff from fundraising, advocacy and programme sections of NGOs and with a number of NGO chief executives. NGOs were selected on the basis of having relatively advanced policy making with regard to the corporate sector and having a record of interaction with the corporate sector.[1] This enabled exploration of aspects of the evolving relationship between NGOs and the corporate sector and of the extent to which a coherent approach in attitudes and strategies was apparent within the development NGO sector.

Development NGO policies on corporate engagement

Most of the NGOs had no single written policy harmonising and coordinating different aspects of engagement with companies. Several larger organisations were in the process of developing, or integrating and upgrading, policies.

Almost all had a policy or code covering fundraising or marketing relationships with companies, and several had a policy or guidelines regarding advocacy and campaigning. These, however, were usually created by different departments within the organisations and were often distinct in style and structure. Advocacy policies usually emphasised the importance of integrity and reputation, warned that financial or operational contact with companies could compromise the effectiveness of advocacy, and pointed to the importance of good preparation of information and research, as well as clarity of objectives and targets.

Marketing or fundraising policies generally centred on what types of company should be avoided as incompatible with the values and identity of the NGO or, in marketing parlance, the 'brand fit'. The simplest contained an exclusion list of business sectors; the more sophisticated, procedures for vetting companies or risk assessment, as well as guidelines on the terms of engagement for financial relationships with companies deemed acceptable for cooperation. Most policies stressed that close engagement was only desirable with companies whose activities were consistent with the NGO's own mission and values.

Some of the NGOs also had policies on ethical investment of their funds and ethical procurement policies for their own organisational purchases of goods and services from companies. Only two had a policy detailing the terms on which the NGO should engage companies for joint programmes or the development of codes of conduct.

NGOs studied identified the potential benefits produced from engagement with the corporate sector. The emphasis, however, varied. Some concentrated on the possibility of fundraising from the sector and for increasing donations from that source. Others emphasised the potential for changing corporate behaviour in pursuit of their mission to benefit the poor. The more radical saw this achieved principally through critical confrontation, while others proposed greater engagement and dialogue.

The process of policy formation

Most NGOs recognised the need to increase corporate engagement; for many, this provided the main impetus for the creation of an overarching policy about relations with the corporate sector. All accepted that relations with the corporate sector cut across several organisational departments and required good internal coordination. In general, the larger, more internationally dispersed and federated NGOs, which faced challenges in effective communication, felt the need for formalised policies related to effective communication and joint decision making. Smaller or single-country NGOs more often felt that individual departmental policies and plans would suffice for the foreseeable future, provided policy formation was accompanied by consultation with other departments and was followed up with good inter-departmental coordination and communication.

Most policy development regarding the corporate sector took place within individual departments, with some wider consultation, or through the work of an inter-departmental working group set up for the purpose. A finished product had then been presented to the organisation's senior management and/or board for discussion, amendment and approval. Several NGOs had formed a 'business group' or 'ambassador's group' from a mixture of NGO staff and individuals from the corporate world. Issues of effective fundraising predominated in their terms of reference and agendas, but they offered more general policy advice about relations with the corporate sector in regard to policy about corporate social responsibility, programme cooperation and advocacy.

Policy position on corporate social responsibility

NGOs tended to adopt very similar positions in terms of analyses of globalisation and the power of TNCs. Most contended that the power of companies had grown and that economic and financial regulation was being reduced by the process of globalisation. Equally, most acknowledged the increasing responsibility of companies for the supply of services vital to the wellbeing of poor people, such as health care, education, water and sanitation. General concern was expressed about corporate influence on governments and inter-governmental bodies, and on the formulation of international trade, investment and financial regulations and agreements.

Although some of the more radical NGOs might take the view that TNCs have no role in a sustainable future (Heap 2000: 17), all the NGOs surveyed accepted that, under the right conditions, TNCs had the potential for positive impact on development. At a more practical level, public messages from development NGOs about what constitutes good corporate behaviour were fairly consistent. Both NGOs which emphasised critical campaign work and those with a more accommodating position towards companies contended that corporate citizenship is not a philanthropic choice but an imperative. Most also acknowledged that the circle of stakeholders to which a company is responsible is not restricted to shareholders, staff and customers, but extends into the communities and countries where companies do business.

NGO analyses of the reasons why companies should be compelled to show greater international responsibility were also fairly consistent, with a high degree of unanimity about the need for companies to eliminate the exploitation of children in their workforce, or in the workforce of their suppliers, and to ensure decent conditions for all employees. The call from NGOs for companies to ensure that essential goods and services, such as pharmaceuticals, were available and affordable by poorer people was similarly universal, as was the call to avoid environmental damage.

This commonality of analysis is illustrated by reports such as *Beyond Philanthropy: The Pharmaceutical Industry, Corporate Social Responsibility and the Developing World*, jointly prepared by Oxfam, Save the Children Fund and VSO (Oxfam *et al.* 2002). For these three NGOs, real corporate responsibility on the part of drug companies would include policies covering their impact in developing countries addressing pricing, patents, joint public–private initiatives, research and development and the appropriate use of medicines

(Oxfam *et al.* 2002: 25). Common views also occur in coalitions such as Publish What You Pay, to which Christian Aid, Save the Children, CAFOD (Catholic Action for Development), Oxfam, World Vision and the Dutch agency NOVIB are all signatories, which called for mandatory disclosure of payments to, and transactions with, governments by multinational extractive companies and their subsidiaries and partners. The intention was to reduce corruption, payments to warring factions, and government mismanagement of state income (Publish What You Pay 2002).

A commonly held analysis of the economic factors contributing to poverty was illustrated by the very broad NGO commitment to the Jubilee 2000 campaign for Third World debt relief at the end of the 1990s. Similarly, there is widespread NGO support for the proposal for a tax on financial transfers around the globe. Often known as the 'Tobin tax', after its initial proponent, the tax was designed to reduce speculation and provide funds for development. The *Declaration on Harnessing Currency Transactions to Tackle Global Poverty* asserts that speculative financial flows contribute to serious economic damage in many countries and calls for a small levy on currency speculation to dampen harmful flows and to raise revenue for poverty alleviation programmes (War on Want 2002a, 2002b). Implementing and managing a global tax require international institutional support. Although that has not materialised, there have been national initiatives in support of the tax, with a modified Tobin tax approved by the Belgian federal parliament in 2004, to be introduced 'once all nations of the eurozone produce a similar law' (Wikipedia 2006b). The Tobin tax has also received support from the presidents of Brazil and Venezuela.

The great majority of NGOs acknowledged the positive impact of corporate activity in terms of contributions to growth, the creation of jobs and the transfer of technology and other business knowledge (Graymore and Bunn 2002). The more analytical NGOs augmented this, noting that corporate activity alone was not the sole condition for poverty reduction. To ensure that corporate activity benefits the majority, it must be accompanied by good governmental and legal systems, adequate laws, a participatory political system and good conduct on the part of companies. Several NGOs gave examples of good corporate conduct, which included engaging in fair competition, complying with the law, and adherence to basic ethical, consumer, labour and environmental standards.

The more critical NGOs maintain that companies can contribute to poverty and inequity unless they are subject to adequate regulation obliging them to conduct themselves in a socially, financially and environmentally responsible manner. Such analysis points to the tendency for larger corporations to use their economic and political influence to destroy smaller-scale local competition, to gain a disproportionate share of access to such assets as land, property, raw materials and markets, and to exert undue political influence locally, nationally and internationally. Using wealth and power to reduce equality of access to productive opportunities, the NGOs argue, has a direct effect on the livelihoods of poor people and exacerbates inequality. Corporations also use their economic power and their promise of large-scale investment to bargain with host governments for looser regulations on labour and environmental protection, for lower taxes and for other special financial incentives such as government investment and subsidies for supporting infrastructure. The NGOs were also critical of companies which move capital and investment rapidly and in a speculative manner from country to country, contributing to financial destabilisation and job insecurity.

A number of NGOs remained quiet on all but the positive aspects of corporate roles, for two distinct reasons. The first was the deliberate emphasis of more moderate development NGOs on fundraising from corporations, with a strategy of not antagonising this source of income through public criticism. Some NGOs also would not or could not maintain an in-depth analysis of the economic and social structures contributing to the existence of poverty sufficient for public campaigning and debate. For this reason, they held no formal position on the role of corporations in development. Less than half of the NGOs were working on a normative statement with specific reference to the role of markets, growth and corporations in the development process, spelling out what constitutes good and bad corporate conduct.

For many NGOs, policy positions about what constitutes responsible behaviour were implicit in a range of materials rather than explicitly laid out in a single positional document. Their views were articulated in advocacy materials about economic issues such as trade or debt, or in campaign reports about certain industries such as the extractive industry or the pharmaceutical industry. Preparatory documents for participation in international conferences such as the World Summit on Sustainable Development, or tripartite initiatives such as Britain's Ethical Trading Initiative and the US Fair Labor Association, contained indications of the NGOs' position on corporations (Graymore and Bunn 2002). In many cases, discussion of corporate conduct was set in the context of calls to governments for more thorough regulation of the corporate sector, rather than addressed directly to the corporations themselves.

There was not always a direct relationship between the content of theoretical papers produced by NGOs and their activities. A World Vision UK analysis of NGO–business partnerships contained some highly critical and sceptical articles about corporations and their cooperation with NGOs, albeit accompanied by a disclaimer noting that the views in the document did not necessarily reflect those of World Vision (World Vision UK 2000). In the document, infant formula campaign activist Patti Rundell questioned the motives of corporations and the value to NGOs of engagement, for 'the rights of TNCs are being enhanced through [NGO–business] relationships while citizens' rights are diminished' (Rundell 2000: 10). A second article suggested that talk of 'partnerships' between businesses and NGOs was often overstated and had become a much abused politically correct development term (Tennyson 2000: 16). Yet World Vision was more generally perceived as an organisation which drew a distinction between 'moderate NGOs' and 'campaigners', and placed itself firmly in the moderate group seeking cooperative and non-confrontational relationships with the corporate sector (Currah 2000: 24–6).

Policies on causes and solutions to poverty

Oxfam Hong Kong noted that underlying any policy on relationships with companies there had to be organisation-wide agreement first about the *type* of social and economic change necessary to tackle the causes of poverty and second about the *process* necessary to achieve that change. Only then could a coherent policy for engagement with the corporate sector follow (Oxfam Hong Kong 2001: 1).

The policy positions about overall solutions to poverty held by the NGOs contained much common ground. All agreed that poverty alleviation required more than welfare handouts, noting that the approach needs to involve 'helping people to help themselves', 'empowerment' or 'providing productive opportunities'. Even the most uncritical of agencies accepted that measures must be taken to increase the control of the poor over resources and that current economic policies tended to benefit those who were already powerful (CARE and WWF 2002: 3).

The fundraising material from several agencies, however, still implied a welfare approach, particularly in the case of child sponsorship, or fundraising material containing itemised lists of 'what your money can buy'. Both portrayed a simplistically direct relationship between the amount of money given by the individual donor and improved wellbeing of individual people or families. This may have had more to do with a simplification of messaging in order to attract donations rather than with an institutional belief in a welfare approach.

Several agencies developed their analysis further, contending that empowering people, or providing productive opportunities, is as much a social process as an economic one and that those who enjoy disproportionate power and wealth seldom cede any to those less powerful without some resistance. Thus, poverty alleviation work can lead to helping people organise and demand their rights or to critical lobbying of governments and corporations.

Less socially active agencies avoided critical confrontation, tending not to extend analysis into this realm, concentrating instead on the positive contribution all parties can make to change. They seldom, however, offered an alternative analysis to the more radical and fundamental critique of economic and social injustice. CARE USA's advocacy manual of 2001, for example, was concerned entirely with influencing governments and inter-governmental organisations. Corporations were not featured in any of the strategies or examples offered (Sprechmann and Pelton 2001).

Encompassing gender issues

Equal opportunity employment was included in the definition of good corporate conduct in a number of NGO policies. Campaigns critical of the marketing of breast milk substitute, joined by many of the NGOs in the past, were clearly of particular relevance to women. Apart from these two items, however, gender issues were not specifically referred to in all but one of the policies surveyed. No attempt was made in any of the analyses of corporations and development to disaggregate the impacts of corporate activity on women and men or to identify or propose specific dimensions of corporate conduct of benefit to women.

NGO staff identified a number of issues of corporate conduct that raised gender issues, although rarely directly. Several felt these should form a part of NGO policies about corporate engagement in terms of lobbying for improvements in corporate conduct, as well as being a feature of the reputational rating of companies (see Box 7.1).

Box 7.1: **Corporate conduct and gender issues**

Corporate conduct which raises gender issues includes:

- the impact of companies hiring migrant labour on the livelihoods of migrants and on their source communities;
- the impact of corporate policies encouraging more flexible employment systems (temporary, part-time and homeworking) and the effect of these on women's economic and social security;
- issues arising from the imposition of codes of conduct by transnational buyers on subcontractors (often as the result of NGO lobbying), such as: a) large, factory-based suppliers being better placed to make the changes necessary to comply with codes of conduct, so other suppliers, such as homeworkers and informal workers (most of whom are women), may be dropped after codes of conduct are imposed; and b) codes of conduct can result in higher operating costs for subcontractors in terms of health and safety provisions, which may be covered by cost savings elsewhere, such as lower wages for workers;
- aggressive marketing of toiletries and cosmetics to women, and the impact on traditional products and practices;
- the portrayal of women in corporate advertising;
- the particular impact on women of the privatisation of basic services, particularly water supply, education and health services;
- aspects of HIV/AIDS drugs and other pharmaceutical distribution policies.

(Source: Interviews with NGO staff)

Corporate relations with the NGO sector

The NGOs saw an influential role for themselves in relation to the impact of corporations on the poor, yet all flagged the danger that partnerships with companies could become successful public relations exercises for companies, with little benefit for either the NGO partner or the poor. The director of one major international NGO network noted, however, that whether or not NGOs which engage with corporations become a 'fig leaf or of genuine added value is determined partly by our own capacity'. If NGOs are not prepared to commit to a professional level of skills and resources when engaging the corporate sector, the value of interaction may indeed become token.

Owing to the lack of homogeneity within NGOs, messages about how they want to work with the corporate sector can become mixed to the point of seeming schizophrenic. This, of course, can be the same on the part of the companies, as decision makers within companies are similarly divided on the value of working with NGOs. In some cases, companies have worked with NGOs and only later find that these NGOs have criticised them elsewhere or suddenly gone cold on further cooperation.

All agencies drew a distinction between those companies from which they would be willing to take funds and those companies with poor reputations which they might consider working with if there was potential and willingness on the part of the company to bring about positive change. There was little evidence of tension among NGOs regarding different approaches to corporate engagement, which were generally a function of the different political complexions of the NGOs. Relations with the corporate sector therefore simply stood as a reinforcement of previously accepted differences of outlook and approach.

An unmet desire for greater coordination between NGOs and corporations was evident. Several NGO staff expressed a need for more clarity about what one called the 'duty of solidarity within the NGO community'. There was certainly sensitivity to the opinion of the NGO community, but no formal activities promoting a more common approach were detected beyond the Corporate Accountability Caucus of like-minded NGOs. Coordination among NGOs on the subject of corporate relations was usually informal, or related to participation in such initiatives as Britain's Ethical Trading Initiative or the Global Reporting Initiative. Although these initiatives involved corporations, academics, government and NGOs, the meetings provided occasions for NGOs to compare notes.

The director of the World Development Movement, Barry Coates, suggested the most important collective challenge facing NGOs was to 'achieve a degree of cohesion, so our combined impacts are greater than our individual activities' (Coates 2000: 30). To facilitate this, NGOs should, on the one hand, avoid undermining one another by inaccurate or sensationalist allegations against companies while, on the other hand, avoiding an approach to engagement that would legitimise or defend companies, where 'the financial or personal gains for individual NGOs can be achieved only at the expense of the overall

Figure 7.1 Nike protest: does this rule out engagement?

movement for change' (Coates 2000: 30). This is the classic tension between private (NGO) gain versus the public (collective) interest.

There was, however, considerable acceptance and tolerance of different approaches to the corporate sector among NGOs. The director of CARE International UK, which was forging a number of marketing and programme links to companies and did not engage in critical campaigns against corporations, said in a 2000 interview: 'I would defend to the hilt NGOs' right to march up and down on the pavement outside head offices and protest against [corporations'] activities' (Hartnell 2000: 32). The policy director of one NGO suggested that greater policy coordination should mean linking and learning, sharing knowledge and best practice, but not necessarily total consistency of approach and tactics. He contended that diversity of approaches by NGOs was valuable, and that 'rock throwers create political space for more constructive engagement by others'. The transaction costs of a unified approach would also be very high in terms of communication and vetting protocols.

The sensitivity of NGOs to one another, even when their analysis differs, is illustrated by the case of World Vision and Nike. World Vision developed a relationship with the US sportswear manufacturer, which donated gifts in kind for World Vision to use in emergencies and development programmes. Nike was subject to a high-profile international campaign over labour conditions in its Asian manufacturing plants, involving several other NGOs around the world, and responded to the critical campaign by introducing a code of conduct in its suppliers' workplaces and other reforms. Criticisms continued, however, with campaigners claiming the changes had not ended labour abuses. Nike also joined a programme called the Global Alliance for Workers and Communities, which aimed to survey the aspirations of its workforce and respond with funded projects. For this, it needed implementing NGOs. World Vision felt that the previous relationship could be the basis for its involvement in these programmes addressing the stated needs of communities affected by Nike's operations. A partnership emerged in which World Vision helped Nike set up and run an education programme for its suppliers' employees in China. World Vision viewed this partnership as a way of using the corporation's presence in the country to address some of the larger social issues that were part of World Vision's mission.

In January 2000, what was then Community Aid Abroad (CAA – now Oxfam Australia), the Australian NGO which coordinated an Australian campaign against Nike, wrote to the president of World Vision to express its concern that the joint programme could reduce the effectiveness of the campaign:

> My concern is that [the corporation] is using its relationship with World Vision to undermine the international campaign to persuade the company to improve conditions in its suppliers' factories. I am writing to request your assistance ensuring that [the corporation] does not use this association with World Vision to mask its problems with labour standards.
>
> (Currah 2000: 26)

While World Vision realised that certain campaigning groups considered the sportswear manufacturer a pariah, it was surprised that others would consider the partnership to be undermining any other campaigns. After internal discussions about whether they should explain the reasoning behind its relationship with a TNC to another NGO, World Vision decided its activities could give the impression that it supported the corporation's public relations work. It accepted that 'the partnership, and World Vision's role in it, could not be isolated from the concerns of other civil society groups' (Currah 2000: 26). World Vision replied to CAA, restating the basis of the partnership and emphasising that this did not mean that World Vision condoned some of the corporation's controversial operations. The continuation of the partnership, it explained, would be on the basis of demonstrable improvement in the company's labour practices, as verified by external groups.

When World Vision informed Nike of the correspondence, the corporation expressed its unhappiness with World Vision's decision to communicate directly with groups campaigning against it. World Vision felt the corporation failed to appreciate 'the pressures on World Vision from its civil society peers to engage with the campaigners'. It was thus caught 'in the middle of a bitter and protracted dispute between campaigners and international business, which its original partnership framework had not accounted for' (Currah 2000: 26). The two NGOs agreed that, in future, the campaigning NGO would share its material with World Vision, which would pass this to staff working on the partnership so that they could be regularly informed of the concerns of campaigners and maintain vigilance with regard to the corporation's activities (Currah 2000: 26).

An example of coordination on policy was the Corporate Accountability Caucus, which came together around the 2002 World Summit on Sustainable Development (WSSD) to push for clearer outcomes on international regulation of corporate accountability. This took the form of an ad hoc coalition, which included Friends of the Earth, Greenpeace International, Christian Aid (UK), Oxfam International, Third World Network and the environmental group FIELD. The coalition pooled ideas about the need for global instruments of corporate accountability, and developed a coordinated strategy for lobbying at the World Summit. High-level sources credit the NGOs with playing a role in ensuring that wording in the summit's implementation agreement included a call for 'urgent action at all levels to: Actively promote corporate responsibility and accountability . . . including through the full development of intergovernmental agreements and measures' (United Nations 2002). This text was threatened with deletion several times during the drafting process, but the NGO coalition lobbied sympathetic government delegations, inter-governmental groupings and influential individuals to ensure its inclusion. The coalition has continued to work together since the WSSD to develop strategies for the application of this part of the implementation plan.

Organisational relations with corporations

Basic approach and working principles

One NGO director commented that 'NGOs need to make it easy for companies to do the right thing, and difficult for companies to do the wrong thing'. In contrast to this, a body of NGOs felt that market-based corporate activity was fundamentally problematic for equitable development. For them, the most effective strategy was to remain independent critics of corporate activity, to have no working links with companies and to campaign for government and international restrictions on corporate activity.

Friends of the Earth, for example, passed a resolution in the late 1990s that the organisation would assume that TNCs have no place in a sustainable future until proven otherwise, and dialogue with TNCs, outside of specific Friends of the Earth campaign needs, would be discouraged. The World Development Movement (WDM) has boycotted invitations to business–NGO forums with companies such as Rio Tinto (Heap 2000: 27, 70). The core of the WDM's

'People before Profits' campaign is a call for governments and international institutions to regulate TNCs (World Development Movement 2002). At the other end of the scale are NGOs that have little hesitation in using the power and resources of companies to support their work. This group has a flexible approach to encouraging companies to support their work, and is unlikely to take a critical position against companies on any issues.

Between these two poles lie the majority of NGOs, which believe that a mixed strategy of advocacy, programme engagement and marketing relationships, designed to fit the nature of the corporation and the issue, will bring the best results. They recognise the problems companies can cause, but believe that the answer is to work for improved corporate practices and to direct corporate power and wealth towards sustainable solutions and beneficial programmes. From a company perspective, however, forming partnerships with NGOs which combine cooperation with criticism is sometimes seen as a high-risk strategy (Heap 2000: 55).

One NGO, although a leader in many critical campaigns against major companies and industries, suggested that companies should still be considered 'innocent until proven guilty' for the purposes of marketing relationships. Some NGOs proposed dual approaches of advocacy and engagement and suggested that the best role for international agencies was brokering or facilitating contacts between companies and local grassroots organisations and communities, rather than implementing any joint work directly with the company. This role enables more direct dialogue and possible joint activity between the company and people affected by the company's activities. It also reduces the risk of conflict of interest, co-option or reputational damage. One NGO queried whether or not the organisation needed a coordinated, proactive strategy of engagement, or simply guidelines and procedures to cover the various less coordinated instances of interaction carried on more or less separately by different departments. Its policy paper opted for the latter approach, and concluded with a series of checklists and communications protocols for use by different departments considering receiving funds from companies, launching critical campaigns, engaging in joint programmes, working on codes of conduct or simply engaging in exploratory dialogue. Many NGO policies noted that relationships with companies should not be undertaken where these might have an adverse impact on any of the pre-existing activities of the agency, such as advocacy or fundraising. This was certainly not the case when advocacy gained prominence in NGOs. It was accepted that some donor income would be affected when NGOs were publicly critical of the status quo.

Oxfam Great Britain noted the importance of practising internally what it advocated externally. The development of a corporate policy was accompanied by the strengthening of the organisation's ethical purchasing policy for goods and services required by the organisation, and an ethical investment policy for the agency's assets (Oxfam Great Britain 1998: 22–3). Oxfam was not alone in this. In some countries, the national NGO council has guidelines to assist member NGOs with ethical purchasing and ethical investment questions.

In several cases, policy documents indicated an awareness of issues, but offered no practical solutions. Most common among these was the question of exiting a programme or marketing relationship with a company when circumstances change. An example of this was the dilemma faced by several NGOs engaged with the oil giant BP. The company was taking serious initiatives on social and environmental impact, but then merged with Amoco of the US, which was a member of the Global Climate Coalition, a US business group lobbying against tighter controls on greenhouse gas emissions, arguing that global warming was an unproven theory. Happily for the NGOs, BP-Amoco also left the Coalition, which in turn disbanded at the end of the 1990s. Another example of difficult changed circumstances came from an NGO with a long-standing marketing relationship with an energy company. The company had a good record on environmental and social concerns but then purchased a coal mine in an Eastern European country with a poor health and safety record in this sector. Shortly after this, the marketing relationship was ended.

Marketing cooperation

Fundraising from corporations represents the longest-standing form of NGO engagement with the corporate sector. Policies covering fundraising from corporations were therefore the first to emerge and the most common among the development NGOs. Marketing or fundraising policies most commonly concentrated on identification of the types of company and sectors of business to be avoided because of their poor reputation and their incompatibility with the purposes and reputation of the NGO. The simplest policies therefore consisted of little more than an exclusionary list of business sectors. More developed policies included procedures for vetting companies, often in the form of a checklist or flowchart. Some also contained guidelines about the terms of engagement for financial relationships with companies which were deemed acceptable for cooperation, but the rules for engagement more often appeared in standardised letters of understanding or contracts entered into by the NGO and the company.

The business sector explicitly, and virtually universally, excluded by development NGOs consisted of companies involved in the manufacture of weapons or their sale to developing countries. This was followed closely by tobacco companies. Most NGOs refused donations from companies producing and marketing breast milk substitute in developing countries. Other NGO policies excluded the receipt of funds from companies known to derive significant amounts of their incomes from extractive industries (oil and mining), pharmaceuticals, alcohol, gambling, nuclear power and uranium production, pesticides and pornography. Several fundraising policies qualified exclusions, using terms such as 'unscrupulous promotion' of certain products or 'aggressive marketing in developing countries'. Many included general, but generally undefined, clauses prohibiting marketing arrangements with companies with 'poor standards' in the areas of labour, environment and human rights. In addition, several NGO policies warned against accepting funding from companies in business sectors against which the organisation was currently holding a critical campaign.

Several NGOs produced materials designed to encourage companies to support them in fundraising partnerships. The main attributes mentioned in NGO material encouraging corporate donations were: their understanding of the needs of business and questions of corporate social responsibility; their experience with on-the-ground development; their favourable public reputation; their values; and their high public profile. Most agencies offered 'customised' fundraising options such as staff fundraising, public events and support for specific projects, countries or types of work.

One NGO noted that the process of vetting companies wishing to donate could in itself become an influencing opportunity, citing the case of a sports shoe company which approached it offering sponsorship for an event. The NGO asked if the company had a code of conduct, a policy about workers rights, or public information about the situation in its plants around the world. The company replied that it did not have these, but was now interested in developing them. The NGO deferred acceptance of sponsorship for a year, hoping to work with the company after it adopted a code of conduct.

For Christian Aid, the basis for marketing decisions was to avoid endorsing activities inimical to the agency's own. This basically meant avoiding endorsing activities that contributed to the creation of poverty. A distinction was drawn between actions contributing to the creation of poverty and actions posing a risk to the name of Christian Aid; their policy saw the latter as a secondary consideration. ActionAid had two fundamental principles about receipt of funds from companies: it would not take money from companies whose products or activities have a direct negative impact on the NGO's beneficiaries, or from companies if, in doing so, the NGO stood to lose other donors. On the other hand, the campaigning organisation World Development Movement has decided that it would generally not take corporate funds as long as it was campaigning against TNCs (Heap 2000: 47–8).

CARE International has had a long history of engagement with the corporate sector for the purposes of fundraising, including a relationship with the coffee chain Starbucks which dates back to 1989. But, in common with most development NGOs, CARE is aware of situations where the activities of a donor might be incompatible with its mission. CARE decided against accepting funds from Monsanto, expressing concern about the impact of marketing genetically modified seeds to poorer farmers, which involved the introduction of complex technology and new recurrent expenses to vulnerable producers (Hartnell 2000: 32).

Critical advocacy, lobbying and campaigning

Advocacy, policy research and public campaigning are the fastest-growing departments of many development NGOs. This stems from a belief that leveraging the policy and practice of governments or international organisations on macro-economic issues such as debt, trade, aid and investment can have more impact on the lives of more poor people than most grant-making activity and community-level development programmes. Advocates point to the fact that changes in macro-economic conditions, such as the East Asian economic crisis of the late 1990s, or political and legislative changes can annul the benefits of decades of village-level community projects.

Cooperation with companies, either as donors or as cooperants in development projects, is viewed with suspicion by many advocacy staff, concerned that this will undermine critical messages and complicate advocacy goals. NGO policies about advocacy tend to emphasise making strategic choices about issues, methodologies and formats for campaign design, rather than strategies for engagement with advocacy targets.

One area of work referred to in some interviews, but not reflected in NGO policies, regards joint lobbying of governments and multilateral bodies with sympathetic and progressive corporations on issues of common interest. Examples of this are scarce, but several interviewees felt that commonality of interest between the more enlightened companies and NGOs was likely to grow and present real opportunities. An example was provided by Greenpeace and the World Business Council for Sustainable Development, which held a joint press conference at the August 2002 World Summit for Sustainable Development. The two former combatants (which both made it clear they might clash again in future) shared a platform to call on participating governments to set aside self-interest and show more commitment in achieving a positive agreement about climate change (Greenpeace 2002).

A commonality of advocacy interests between companies and NGOs can involve national and local companies. Where NGOs are lobbying for fairer prices or better export market access for items such as agricultural commodities from developing countries, there is a clear opportunity for a beneficial alliance with local or national producer companies. One example was Honduran coffee producers, some of them quite large companies, which added their support to the Coffee Rescue Plan proposed by NGOs, calling for measures to increase and stabilise coffee prices.

Programme cooperation

Programme cooperation is a new area of corporate engagement, which does not completely fit with current programmatic skills of providing grants and other support to grassroots partner organisations. Where NGO policies do mention operational engagement with companies, this is mostly concerned with international work to change policy and practice at headquarters level. However, as outlined in Chapter 6, companies are increasing their offers of finance for specific projects related to their developing country operations, and also developing a range of project work of their own, requesting partnerships with NGOs operating locally. Programme staff in the front line are not prepared or trained to deal with these corporate approaches, nor do most organisational policies provide guidance for doing so. For NGO staff, analytical skills relevant to the range of issues and problems related to corporate activity need to be developed before training about forms of engagement with corporations can be effective.

The UK director of one NGO argued that all NGOs need to develop the capacity to engage when interesting approaches come from companies. His organisation had been approached by companies undertaking major mining developments in an African country wanting help with assessing the social impact of the project, and by another company looking to improve conditions in its supply chain in South-East Asia. In both cases, local staff capacity was insufficient, although local staff saw the potential impact they could have and were wholeheartedly in favour of engagement.

The reaction of country-level programme staff to approaches from corporations seeking cooperation was varied. At one extreme was total refusal to engage with companies under any circumstances on ideological grounds. At the opposite extreme was the tendency of some field staff to agree too easily to offers of financial support from companies, without considering the reputation of the company. Other reactions included indifference, often an indication of an already overly full work schedule.

Some NGO staff noted that awareness of the issue of corporate engagement, and skills in negotiating joint work, might not always revolve around the NGO's own engagement. On occasion, country-level staff had received requests for advice from local partner organisations which had been approached by corporations with offers of money and cooperation. As with other forms of programme cooperation, training is clearly needed for the NGOs to be able to play an effective role in facilitating contact and mediating relationships. Some models for such a role exist. ActionAid's India operation became interested in the idea of corporate partnerships for development in the early 1990s. The organisation took an original approach by setting up an 'arm's length' organisation to foster partnerships between the corporate sector and social development initiatives aimed at the eradication of poverty (ActionAid 2000: 2). The resulting organisation, Partners in Change, considered that its mission was to 'increase the understanding and active participation of business in equitable social development as an integral part of good business practice by promoting partnerships between business, disadvantaged communities, development initiatives and government' (Partners in Change 2002). In this, the organisation set its goals as: increasing understanding in the business world of the need for operating in a socially responsible manner; mainstreaming socially responsible behaviour in business, based on an understanding of the impact of business on disadvantaged communities; creating an enabling environment for building partnerships between business, civil society and the state which benefit communities; helping an increasing number of businesses become partners in sustainable development programmes; and developing monitoring tools for the impact of these corporate programmes on the lives of the disadvantaged (Partners in Change 2002).

Partners in Change played no role in helping NGOs raise funds from companies, but attempted to build partnerships between NGOs and companies which it believed would help mainstream responsible behaviour amongst businesses. Its work centred on: research and recording of best practice with a view to creating training materials for business, NGOs and government; assisting the creation of corporate policies and benchmarks on social responsibility; carrying out training for corporate, NGO and government staff to build capacity for effective partnerships; facilitating partnership formation and dialogue; and assessing the impact of business–NGO–government partnerships (Partners in Change 2002). Partners in Change developed a comprehensive strategy for involving corporations in the development process, and gained a great deal of experience from the many corporate–NGO projects it helped facilitate.

For several interviewees, NGOs should engage with national and local businesses, small and medium-sized enterprises, cooperatives, and worker-owned businesses at a programmatic level, rather than investing time and energy in engaging giant TNCs. Others saw these two sectors as distinct in terms of NGO strategies. In the case of cooperatives and smaller, local enterprises, NGOs wanted to see them grow and prosper, as well as improve their conduct, which might involve offering programme support. In the case of TNCs, on the other hand, the aim was to alter policies and practices to improve their impact.

Dialogue and development of standards

While many NGOs have participated in the development of codes of conduct and industry standards, little in-depth strategy for this was evident in any of their policies. NGO materials about this subject reflect a growing scepticism. Having been quite active in the development of codes of conduct in the mid- to late 1990s, many NGOs subsequently criticised the superficial application of such codes. Other criticisms suggested that many codes were designed in a way most appropriate for unionised male workers, and failed to reflect the family responsibilities and reproductive needs of unorganised female workers (Dhanarajan 2000: 7–8).

On the other hand, NGOs credited codes with sensitising management to the problems workers may face and with providing workers with tools for strengthening their rights and for making interim improvements in their situation, even where they were not free to organise (Dhanarajan 2000: 9). Programme staff in one NGO, for example, indicated that they had been approached by local employers, including textile factory owners in Bangladesh, requesting advice about priority issues in employment, health and safety. The factory owners were motivated by awareness that several of their major customers were adopting codes of conduct for suppliers and they wished to be better prepared.

While NGO enthusiasm for individual company codes of conduct has waned, several seemed more willing to remain in broader multi-sectoral initiatives related to corporate conduct and ethical trading. Several were involved with the

Global Reporting Initiative – an industry, academic and NGO initiative to set reporting standards for corporations on environmental, labour, human rights, financial and social impact. Several British NGOs were involved with the Ethical Trading Initiative, a tripartite alliance which seeks to set labour standards for investment and supply chain activities in developing countries. In the few cases where NGO policies did mention work about voluntary codes and standards, they were explicit that this work should only be undertaken on the clear understanding that voluntary codes and standards were not a substitute for more binding national and international regulatory mechanisms governing corporate accountability and responsibility.

Several NGOs also promoted fair trade standards and goods. Such fair trade standards differ from corporate codes of conduct in that fair trade goods are priced and traded with the primary intention of ensuring that poorer producers and workers receive more for their products and that income is invested in improving the community and reducing poverty (Heap 2000: 118–19). Involvement of NGOs in this type of work involves funding programmes to assist producers in developing countries with training, start-up costs or infrastructural development, or with lobbying retailers in developed countries to sell more fair trade goods.

Table 7.1 illustrates the range of NGO engagement across a number of business sectors, and was compiled from material gathered from interviews and literature searches. Most of the issues noted are the concern of more than one NGO.

The risks of engagement

NGO policies emphasised the importance of protecting NGOs' reputation against criticisms that their activities were being compromised by engaging in financial or working relationships with companies. For the more publicly critical NGOs, the advocacy department's concerns predominated and policies generally asserted that principled stands on policy should take primacy over financial considerations, in order to ensure that the organisation's integrity and reputation were maintained.

In those NGOs with a strong advocacy capacity, policy and advocacy departments were more frequently called upon to prepare organisational policy than the marketing divisions. These policies betray their departmental origin. One major British NGO's policy, for example, was written by the policy director. It proposed that, in event of a conflict of interest in engagement with a company, 'the primary driver of [the agency's] corporate engagement will be to influence the policies and behaviour of companies to bring about concrete gains for poor people'. The document concluded that corporate engagement should be centrally coordinated, and that: 'In view of the primacy of influencing corporate behaviour . . . the Policy Director will lead in co-ordinating the development and implementation of [the agency's] corporate engagement strategy. . . .' The resulting policies tended to emphasise the risk to credibility

Table 7.1 NGO–corporate interaction by sector

Corporate sector	Advocacy and campaign issues	Dialogue and development of standards	Funding cooperation	Programme cooperation
Aerospace	Weapons production and sales to developing countries			
Agriculture	Dumping subsidised foods; support for trade restrictions; union busting	Quality and organic standards		Community-based agricultural projects
Automobiles	Weapons production and sales to developing countries			
Chemicals	Environment; health and safety (especially asbestos); animal testing			
Energy, hydropower dams, engineering	Environment; land rights; resettlement; inappropriate infrastructure development			Joint projects in communities affected by operations
Extractive (oil, minerals, timber)	War economies (payments to combatants); land rights and access; environment; social, economic and human rights impact on indigenous people	Codes of conduct; multi-sectoral forums; dialogue with affected communities; Forest Stewardship Council standards		Joint projects in communities affected by operations
Finance and banking	Debt; tax havens; speculative capital flows; loans to repressive governments; financing of inappropriate infra-structure projects; financing the arms trade	Ethical investment criteria and monitoring	Credit cards; pro bono accountancy services	Finance for micro-credit programmes

Fisheries	Fishing rights, quotas and practices harmful to indigenous fishing communities	Marine Stewardship Council standards		
Food processors and retailers	Low prices paid to producers; harmful marketing of infant formula	Fair trade standards; marketing standards; quality and organic standards		Joint community development and training programmes with agricultural suppliers
Furniture, home furnishing, do-it-yourself	Child labour; poor health and safety; deforestation	'Rugmark' child labour carpet standard; Forest Stewardship Council timber standards		Education projects with child workers
Information and communications technology	Weapons systems sales to the developing world; pricing widening the digital divide; monopolisation		Affinity internet access services; 'donation for every visit' websites; concessionary supplies of hardware, software and IT services	Communications infrastructure and technical staff in emergency situations; joint education projects
Light industry (shoes, toys, garments, sports goods)	Poor health and safety; child labour; labour rights	Codes of conduct development and monitoring; Fair Labor Association (US); Ethical Trading Initiative (UK); labour standards		'Post employment' training of factory workers; health and safety training; education projects for child workers
Management consultants			Pro bono organisational reviews; training	
Pharmaceuticals, toiletries and cosmetics	Lack of affordable access to medicines by poor people (especially HIV/AIDS drugs); restrictive patents and intellectual property; research bias; animal testing		Concessionary and free supplies of pharmaceuticals	Health programmes; clinics

Table 7.1 (continued)

Corporate sector	Advocacy and campaign issues	Dialogue and development of standards	Funding cooperation	Programme cooperation
Seed, biotechnology and agrochemical producers	Restrictive seed and plant patents (TRIPS); genetically modified seeds; cost of inputs for poor farmers; recurrent dependency on inputs; animal testing			
Tobacco	Aggressive marketing in developing countries			
Travel and tourism	Environment; child prostitution; labour rights	Environmental and development NGOs working with the industry on an accreditation body or stewardship council	Concessionary air fares	
General	Tax avoidance; investments in countries with repressive governments (e.g. Myanmar, apartheid South Africa, Pinochet's Chile); poor corporate HIV/AIDS policies; lack of inter-national regulation of transnational business	Global Reporting Initiative; Global Compact	A range of donation, marketing, co-branding and sponsorship relationships; staff volunteer schemes	A range of project and programme linkages

Note:
This table was compiled from internal and public documents and staff interviews from the following agencies: ActionAid; CARE USA; Christian Aid; Médecins san Frontières Belgium; Oxfam America; Oxfam Community Aid Abroad; Oxfam International; Oxfam GB; Save the Children Alliance; Save the Children Fund UK; Save the Children USA; World Vision UK; World Vision USA and the World Development Movement.

of the NGO's voice caused by its forming financial relationships with companies, and the policing and regulating of corporate links, rather than strategies that sought opportunities to leverage influence or resources through cooperation. Few policies drafted by advocacy departments raised the risk of reputational danger from being overly critical, or through being over-zealous, or engaging in misleading or misjudged attacks, but many were concerned about being seen as too accommodating.

Those NGOs placing greater emphasis on fundraising from corporations and less on advocacy activity also flagged concern that financial association with corporations might damage their reputation; they generally had processes for checking out the reputation of companies prior to any financial tie-up. One interviewee mentioned a further form of reputational risk created by turning down offers of money on ethical grounds, citing the case of his organisation refusing money from a source considered to be incompatible with the principles of the organisation. When this potential donor publicised the NGO's refusal, the public reaction was critical, questioning whether a development agency should turn down 'money which could be put to use helping poor people' on the grounds of esoteric political principles.

NGO policies contained considerably less guidance about the direct financial risks of critical advocacy towards companies. Policies which did list such potential risks saw them in terms of being sued for libel, being hit by restraining orders, having bank accounts frozen, facing security risks to staff, particularly in developing countries, or being subjected to corporate PR counter-attacks. In the United States, lawsuits against NGOs, particularly in the environmental sector, have become sufficiently frequent to be given the term 'strategic lawsuits against public participation'. Some have argued that companies will increasingly resort to such measures (Monbiot 1997: 25). Some NGOs felt their policies did not emphasise sufficiently the risk of legal reprisal. Others felt that NGOs had sufficiently good reputations and high levels of public trust for corporations to realise that suing them was not a realistic option. Companies know that any court case against an NGO would probably only compound any accusations already made against them and add to bad publicity. Respondents cited the 'McLibel' trial, in which the fast food company McDonald's sued two environmental activists from London Greenpeace, winning the legal battle, but costing millions and generating only bad publicity for the company.

It is very hard to ascertain whether or not NGOs which emphasise fundraising over critical advocacy weigh up the financial implications of any criticism of corporations that might be potential donors. Indeed, the decision is unlikely ever to be articulated in such terms. One federated NGO revealed that critical advocacy plans were shelved when it was discovered that a member organisation had a pre-existing fundraising relationship with the company intended for criticism, because of a perceived risk of being viewed as hypocritical or inconsistent. Several NGOs noted that public donor reaction, as separate from corporate donor reaction, was carefully weighed up when public campaigns critical of governments or companies were being considered.

Figure 7.2 People before profits: protestors outside the high court in Pretoria demand affordable AIDS drugs

Some NGO staff suggested there was a risk that the costs of engaging with companies would become so high that more benefits would be gained from alternative activities. Diminishing returns from complex and time-consuming engagement with a company reduce the impact. The transaction costs of effective vetting of corporate donors, and the preparation of credible advocacy efforts, can be very high.

Internal tensions

Concern that critical advocacy should have primacy over fundraising or joint programme work was summed up in the words of one NGO staff member: 'we should be concerned with how profits are made rather than how they are spent'. The most common tension in NGOs lies between, on the one hand, the imperatives of the advocacy and campaigning departments to maintain a consistent and credible reputation for independent, value-based criticism of government and business policy and, on the other hand, the obligation of the marketing or fundraising staff to raise sufficient income for the organisation to continue to function. Fundraisers are often under pressure to look for new methods and new sources of funds, including from governmental and business sources. Many marketing departments saw unfulfilled income potential from corporate sources as increasingly attractive in times of government aid cutbacks and 'donor fatigue' or at least fickleness on the part of the public. Fundraisers, therefore, have the weakest authority in terms of organisational values and principles, but a great deal of legitimacy in practical terms of organisational survival.

The discord between marketing and advocacy departments was best summed up in the words of one marketing director, who said that each agency should 'recognise the inter-departmental tensions, and learn to work through these on an issue-by-issue basis. They will never be reconciled.' This, however, came from an agency with one of the more developed and integrated policies about corporate relationships.

In general, NGOs acknowledged the primacy of organisational values, integrity and reputation above all else. In practical terms, this meant that, where there was a conflict between raising income from a company and the impact the company might have on the message and reputation of the NGO, reputational considerations prevailed and funds were not accepted. Stories abound of funds turned down because a company's activities were considered inconsistent with the principles and beliefs of the NGOs, or because of its potential to compromise present or future advocacy positions. There were far fewer cases of criticism of corporations being muted to ensure that income was not affected. The more moderate NGOs, however, made virtually no critical public reference to the corporate sector, whether from principle or from pragmatism. However, critical advocacy about international issues was, in some cases, muted, such as that about Israel–Palestine issues, because of fears of reaction of significant donors. Critical advocacy could sometimes be curtailed owing to fears of the impact on public donors rather than corporate donors.

In the case of some agencies, the importance of increasing income seems to have dominated policy making. In these cases, the primary initiative for approaching companies lies with the fundraisers. The onus will be on policy or programme staff to make the case that any donation or marketing plan runs counter to the values or principles of the organisation.

The potential for tension between advocacy and programme departments is harder to resolve in terms of values and principles. In many NGOs, programme staff accuse advocacy staff of not grounding their work in real issues facing poor people, with whom staff in the field have contact on a daily basis. The advocacy response is that they are working on macro-economic issues which underlie the structural causes of poverty. In this debate, the programme department can support their case with the opinions of local partner organisations which are influential within NGOs. This is significant since an important component of the legitimacy of international NGOs when they voice an opinion is that they are more in touch with, and aligned with, local grassroots and activist opinion than are large multilateral agencies like the World Bank, governments or companies.

Conclusion

This chapter has examined ideas and observations gathered from a broad cross-section of development NGOs, to present an aggregate picture of the principles, policies and practices which govern their relationships with companies. It revealed substantial diversity within the NGO community with regard to policies about practical engagement with companies, in terms of both their attitudes to companies and the degree to which the policies cover different types of relationships with companies. At the level of practical cooperation, NGOs exhibit very different approaches to corporations. Among those NGOs which emphasise public advocacy and campaigns, however, the approach is fairly consistent.

Unlike the trade union movement, which has the structures and mechanisms to agree to common global positions about issues, the NGO movement is diverse in almost every dimension, including goals, principles and approach. There are therefore dangers that companies can play NGOs off against one another. By engaging the more compliant NGOs in cooperative projects, or in the development and monitoring of codes and standards, they can diffuse, deflect or confuse the more fundamental criticism from the more critical NGOs about their impact on development, the environment or human rights. However, a number of factors work against opportunities for companies to adopt a 'divide and rule' strategy towards NGOs:

1   There is a level of consistency of analysis among those NGOs which do speak out in advocacy campaigns concerning what constitutes good or bad corporate conduct. This makes it harder for companies to evade more general criticisms of their social or environmental impact by co-opting compliant NGOs.

2   Even those NGOs which are unwilling to criticise corporations are sensitive to the position of the campaigning NGOs which are critical of corporations. They are therefore willing to moderate their interaction with corporations which are subject to criticism by other NGOs out of a sense of allegiance to civil society as a whole.

3   The more critical NGOs retain substantial capacity to publicise hypocrisy and invalidate the investment a company may have made in superficial corporate citizenship work, even where this involves partnership with other NGOs.

The lesson for the corporate sector is that the soundest investment in social responsibility will be that which is fundamental, honest and sustainable, representing an effective answer to the accusations of the harshest critics and enabling genuine change in people's lives.

Many corporations are increasing the resources they are willing to offer for philanthropic donations, which they view as an important demonstration of their growing social responsibility. A reduction in governmental aid and fickle public behaviour in charitable donations make corporations an increasingly attractive source of income for development NGOs. If the proportion of income NGOs receive from corporations increases, the importance of coherent policies analysing the role of corporations in development and setting out a strategy for engagement will grow. Ad hoc policy decisions will not protect NGOs from inconsistencies that will challenge their reputation and credibility or pose other types of risk.

Companies are increasingly aware of the need to operate in a way that demonstrates social responsibility, especially where their activities expand in areas with direct impact on poor people. At the same time, whatever their strategy, NGOs all agreed their interaction with the corporate sector could bring about change to corporate behaviour in a way that would have a positive impact on poverty. However, current NGO policy making about relations with the corporate sector emphasises ways to achieve consistency and reduce risk for their traditional activities of critical advocacy and marketing. NGO policies are, in general, failing to point the way to new types of activities with companies and to opportunities to achieve new levels of influence and change.

NGOs' policies have been designed to be most effective at ensuring they do not rush into relationships with companies which might compromise their integrity. The contemporary challenge for NGOs is to develop policies which are sophisticated enough to ensure their integrity and reputation, whilst taking better advantage of the serious opportunities which exist in partnership with the corporate sector. New ways of working with corporations will require some boldness and, as with all things new, bear some risk. Those involved in development work admit that current efforts are inadequate to reduce the number of people suffering illness and death each day from poverty-related causes. Business as usual is not enough, either in the world of business or in the work of NGOs.

Part IV

Dam(n)ing the Mekong?

Banks, states, NGOs and the poor

8 Advocacy, civil society and the state in the Mekong Region

Philip Hirsch

There is little awareness, and even less systematic study, of the shift by local NGOs in developing countries from supporting small-scale community development towards advocacy for policy change. NGO advocacy for better development in poorer countries continues to be understood largely as a 'First World' or global phenomenon. This chapter explores developing country advocacy within South-East Asia, and specifically within the region known as the Mekong, with particular reference to Thailand.

International NGOs have mobilised to pressure OECD country governments to spend more money on foreign aid, specifically to achieve the target of 0.7 per cent of GDP. They have increasingly also targeted donor governments and multilateral development banks and international development agencies to practise development in more socially equitable and environmentally sustainable ways. NGO advocacy is thus commonly understood as a political dialogue between civil society, government and international agencies in a rich-country context. In part, the critical advocacy agenda is informed by 'post-development' thinking, a response to development failures and disillusionment with the modernisation/development ideal and its ability to improve the lot of the world's poor, or to achieve economic wellbeing without destroying the natural environment.

Thus, the association of development advocacy with funding issues and with post-development thinking has tended to link it with rich-country agendas. Yet critical approaches to development are increasingly entrenched within developing societies. Post-development and other critical approaches are partly a product of international currents in the academic world, but partly also emanate from connections and movements within and between international NGOs and their national and local partners in aid recipient countries, problematic though such relationships may be (Jordan and van Tuijl 2000). Networks have become an empowering aspect of rights-based advocacy within developing countries, globally and in specific regions (Perreault 2003).

For the most part, however, endogenous experiences and socio-political processes help explain development advocacy 'from within' in poorer or middle-income countries. An 'upscaling' (Uvin and Brown 2000) of local, small-scale action that has been the traditional preserve and strength of NGOs has taken them into the policy arena (Annis 1987). Further, NGOs are having to respond to rapidly changing society–state relations as neoliberal influences and governance agendas redefine the state's role vis-à-vis the market and open up new spaces for civil society (Bebbington 1997; Fowler 2000a). There are also causes for NGO advocacy specific to developing countries, for example the critique of 'Northern' protected area approaches to conservation that territorialise (Vandergeest 1996) and alienate rural communities from their resource base (Roper 2000).

A cursory review of the literature on civil society, advocacy and development suggests that a higher level of examination within developing countries is needed to understand advocacy beyond global and first-world organisational experience. In some cases, NGOs have become increasingly enmeshed in local and national governance processes from which they had previously been excluded or that they had themselves shunned (Felisa 2004). Shigetomi (2002) gives a rich review of NGOs within developing and developed countries of Asia, showing how movements have emerged in response to quite specific politico-historical contingencies and continue to be constrained and shaped by such contexts, notably by the economic and political space afforded to them or carved out by them.

Context is all-important in defining and shaping advocacy at a national level. The degree of political space in which advocacy may be conducted and the organisational form that it takes vary significantly from one country to another. Variations in developing country experience of advocacy can be explained by a range of factors. On the one hand, trajectories of advocacy can be seen both as countries experience development and its various impacts and as political space emerges in part as a product of the development process. Moreover, as countries move to 'beyond aid' levels of development, in-country NGOs need to redefine themselves with regard to domestic constituencies and, in part, this enhances their advocacy role (Aldaba *et al.* 2000; Fowler 2000b). On the other hand, there is no linear or unidirectional experience. There are characteristic 'world region' differences, for example between Latin America (Foweraker 2001), South Asia (Haque 2002) and Africa (Mohan 2002). However, even within particular regions, there are sharp differences depending on historical and current socio-political configurations. Moreover, space for advocacy waxes and wanes within particular countries.

This chapter examines the differentiated landscape of one part of South-East Asia, based on different levels of development, degrees of political space, cultures of critical thinking about development and other variables. A case study of Thailand shows how NGO advocacy has emerged in the country with the most critical social movements, and shows how the role of NGOs has shifted in response to that country's social, political and economic development. Contrasts are then drawn with neighbouring countries where NGO advocacy is much more circumscribed. The chapter concludes by addressing the vexed question of a second-generation 'upscaling' of advocacy from within: the spilling across borders of an NGO movement able to deal with development agendas at a regional level.

Advocacy in the Mekong: a differentiated landscape

The Mekong is defined both as a trans-boundary river basin and as a broader development region. Advocacy agendas are shaped by these two distinct but related regional formulations.

As a river basin, the Mekong is defined in bio-geographical terms. The bioregion covers a territory of 795,000 square kilometres and has a population of some 70 million people, more than 80 per cent of whom live in rural areas and depend on the basin's resources for their livelihoods. The river flows through, and is shared by, six countries: China, Burma (Myanmar), the Lao People's Democratic Republic (Laos), Thailand, Cambodia and Vietnam. It is also shared by people from different ethnic groups, by farmers, fishers and those with development designs on the river. The main advocacy issues have to do with sustainable development of the river and the impacts that large-scale infrastructural development has on the livelihoods of subsistence-oriented users of the basin's land, water, fish and forest resources.

As a development region at the heart of some of the world's most rapidly growing economies, the Mekong is defined through the Asian Development Bank's Greater Mekong Subregion (GMS) formulation of a zone previously held back by conflict and now ripe for development through large-scale investment. Roads, railways, telecommunications, dams and an associated electricity grid, tourism projects and other physical linkages will bring about this development vision. The main advocacy issues here concern the marginalising impacts of an infrastructure-oriented development programme. Chapter 9 deals specifically with NGO encounters with this programme.

The Mekong Region has become iconic in discussions about development (Kaosa-ard and Dore 2003; Osborne 2004). In part, this is because of the metaphoric significance of the Mekong, the value of which lies in the riverine sense of linkage and flows. The Mekong as a development region is defined largely by the manner in which evolving physical, economic and other linkages are breaking down old barriers and instigating flows of goods and financial capital. Another reason that the Mekong has become the subject of so much attention is that it has all the ingredients for analysis and critique of developmental paths and their impacts. Thailand's long experience with development and the social and environmental depredations resulting from that experience give both a model and a warning to neighbouring countries. In turn, these neighbours were long held back from fast-track, market based, outward-looking development by war and by experimentation with centrally planned socialist economic models that have now been discredited and largely discarded. To what extent will Vietnam, Laos and Cambodia follow Thailand's path and engender the kind of home-grown advocacy seen in Thailand? To what extent do Thailand's own resource needs exacerbate or drive the fast-track, resource-

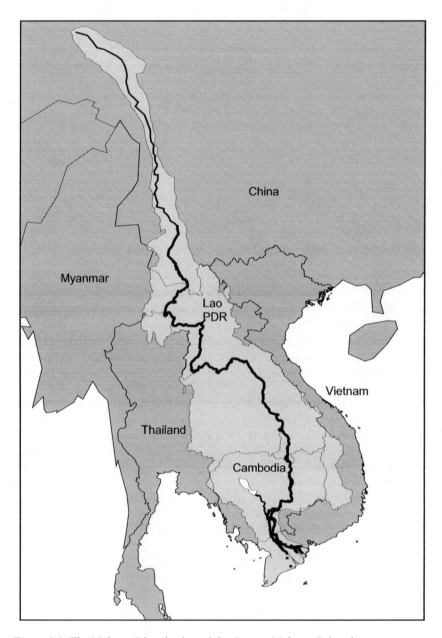

Figure 8.1 The Mekong River basin and the Greater Mekong Subregion

oriented path of development in its neighbours and thus turn Thailand's own advocacy movement toward a regional frame of reference? What are the implications of China's extension of its own breakneck developmentalism into the part of the Mekong within its own territory (Yunnan Province) and further into its integration with the wider Mekong Region, and where does that leave advocates for just and sustainable development with regard to a country not historically open to activist influence from without or within?

While advocacy agendas related to the Mekong basin and development region as a whole can be considered, most policy advocacy occurs at a national level. The Mekong contains a fragmented and diverse political landscape for advocacy. On the one hand, Thailand has seen the emergence of a large and vibrant NGO movement, albeit one that has become increasingly fragmented, internally differentiated and subject to ebbs and flows in the political space and influence that it enjoys. Thailand has also seen the emergence of civil society forms beyond NGOs. At the other extreme, Laos and Myanmar proscribe local NGOs. Vietnam provides some space for non-government organisations, but they are quite different entities from those in Thailand. China has seen a burgeoning of NGO activity, but largely focused on environmental issues. Cambodia has its own increasingly active advocacy NGOs (Hirsch 2001).

Superimposed on all of this is an international NGO presence that is closely intertwined with advocacy within each country. Advocacy is both at individual country level and also oriented vis-à-vis regional institutions such as the Mekong River Commission, Greater Mekong Subregion and, to a lesser extent, ASEAN. This complex regional landscape will now be considered.

Thailand: from community development to advocacy

NGO advocacy in Thailand currently occurs in a society keenly aware of the social and environmental impacts of fast-track development. It also occurs in a relatively unobstructed political space. This space is itself a product of recent histories of struggle, confrontation and accommodation. Civil society and associated advocacy in Thailand has emerged with NGOs, it has bolstered NGOs, but it is not and never has been entirely limited to NGOs.

The roots of Thailand's NGO movement are in the Thai Rural Reconstruction Movement (TRRM), a national branch of the International Rural Reconstruction Movement (IRRM) founded in the Philippines by James C. Yen. During the late 1960s, TRRM operated within the confines of a dictatorial military regime and was largely oriented to supporting community development to address the problems of Thailand's rural poor. Key individuals included Dr Puey Ungpakorn, governor of the Bank of Thailand and later rector of Thammasat University, who recognised early on that the country's economic development was leaving the rural poor far behind. TRRM, and other rural development NGOs spawned from it, steered clear of direct advocacy, but they emerged at a time when Bangkok-based students were starting to get a sense of the gross inequalities and injustices in their country. This realisation was both behind and facilitated by a growing leftist influence and opposition to the US-supported right-wing dictatorship of Thanom, Praphas and Narong.

The sharpening left–right divisions of the late 1960s and early 1970s were exacerbated by events in neighbouring Indo-China. However, NGOs mainly kept their heads down during this period, despite the emancipatory ideals of key figures. At this stage in Thailand's development history, it was the left-led student movement and urban workers who effected change through the 14 October 1973 uprising. A series of escalating events associated with corruption and abuses of power by the right-wing military dictatorship led to demand for constitutional reform, to mounting and increasingly frequent demonstrations, and finally to a violent response from the regime on the streets of Bangkok. Ultimately, this led to a successful overthrow of the dictatorship.

The relatively open democratic period between 1973 and 1976 saw an accelerating awareness of rural injustice as students spent extended periods in the countryside. While a number of NGOs emerged at this time, for example the Komol Keemthong Foundation dealing with urban slum issues and the Union for Civil Liberties dealing with human rights issues, in the public mind the role of NGOs remained that of charitable community development organisations. Political advocacy occurred largely through demonstrations, direct action and university student-based pressure for policy reform in areas such as land tenure. A military crackdown in 1976 resulted in political polarisation, complete loss of political space for independent challenge and a period of intensified armed struggle as students and workers fled the cities to join the Communist Party of Thailand in the forests and countryside.

These seeds of Thailand's current NGO movement sprouted from the early 1980s onward. In 1980, the military made a key decision (in policy order 66/23) to win over the countryside through political rather than solely military means, and rural development played a prominent role in this hearts-and-minds approach. During this time, a government amnesty built on divisions within the leftist movement resulted in a return to the cities of many who had fled in 1976. Nevertheless, political limits placed on challenges to authority about land and about other issues with a leftist association meant that NGOs concentrated their efforts on assisting the rural poor in livelihood development and village-based service provision – priorities that were in tune with the new government policy order.

It was the environmental movement that allowed the emergence of a more advocacy-based NGO approach. Without the political resonance of land- and livelihood-based advocacy, environmental advocacy brought together diverse social forces. It did so under an inclusive and legitimising discourse of concern for the country's declining ecological health and rapidly depleting natural resources, which had resonance at elite as well as grassroots levels. Moreover, significant early achievements, including the cancellation of the Nam Choan Dam and the revoking of logging concessions in 1988 and 1989 respectively, gave the movement a growing confidence. This was further entrenched, and enlisted further elite support, with events following the suicide of Seub

Nakhasathien, a committed environmentalist, government officer and manager of the Huai Kha Khaeng Wildlife Sanctuary. Seub's suicide was interpreted as a cry against the high-level corruption and local pressures that were decimating the country's remaining wildlife, and a host of NGOs and other organisations combined to establish the Seub Nakhasathien Foundation (Hirsch 1993).

The subsequent blossoming of environmental NGOs during the 1990s also saw a growing divide between those which were more purely conservation-oriented and more urban and elite-based, and those whose environmental concerns had their roots in social justice and rural livelihood issues. Both sets of NGOs found increasing space in the policy arena, and advocacy about national park issues, the Community Forest Bill and other key areas became a central part of their work. Some of these NGOs continued to work in local environmental restoration projects, but even in practical areas such as community forestry there was a significant advocacy role that became part and parcel of NGO identity and modus operandi (Hirsch 1998).

The emergence of livelihood-based environmental NGOs, such as the Project for Ecological Recovery, engendered a growing confidence among NGOs to become involved in other advocacy areas related to rights and livelihood. The Union for Civil Liberties took on human rights issues around such areas as land alienation and the infringement of worker rights. In northern Thailand, a growing number of NGOs concerned with citizenship and land rights of ethnic minorities emerged during the 1990s. NGOs working in the HIV/AIDS arena matched their educational and health support programmes with vigorous campaigning about the rights of people living with the virus and urged the government out of its initial reluctance to publicise the problem for fear of offending the tourist industry.

The spaces for NGOs improved markedly during the 1990s. However, the path was not unidirectional. The coup d'état of February 1991 and the military crackdown on protestors in May 1992 saw a severe but short-lived curtailing of civil action. Ironically, the appointed government of Anan Panyarachun was quite favourable to NGOs as a progressive force in Thai society and, with the return to elected democracy, there was a rapid increase in mobilisation. Key issues took hold of the NGO agenda, including opposition to construction of large dams and mobilisation for compensation of villagers affected by those that had been built in north-eastern Thailand (notably Pak Mun and Rasi Salai). However, new forms of civil society action also emerged in parallel; for example, the Forum of the Poor (Missingham 2003) was founded, and provincial civil society associations (*prachakhom changwad*) emerged in several key provinces such as Nan, Phuket and Kanchanaburi. The role of NGOs shifted somewhat in response to this more grassroots and locally-based mobilisation, taking on a support and networking role. Despite the political space afforded by more full-blown democracy, particularly after promulgation of a progressive new constitution in 1997, NGOs at this time were also facing their own challenges, including loss of international funding as donors shifted their attention away from Thailand toward neighbouring countries in Indo-China.

Figure 8.2 The Mun River: a mural reflecting ways of life prior to construction of the Pak Mun Dam

New roles and challenges for NGOs arose with the Thaksin government. Thaksin recruited a number of high-profile NGO workers to the Thai Rak Thai Party, which took power in 2001 (Phongpaichit and Baker 2004). On Thaksin's part, this was a canny move to mobilise – and partially co-opt – NGO support in key localities. In turn, the NGO workers were attracted by the opportunity to influence policy at the highest level from within rather than by advocating from without. Thaksin's populist measures, including cheap health care, village development funds and support for village-level enterprise, provided a justification for those who joined the government. It also resulted in considerable acrimony, which only grew as Thaksin's popularity waned and his monopolistic economic and political practices became more apparent. Ironically, by the time of the elections in April 2006, NGOs had come full circle into the forefront as the only force – combined with academia and some media and business competitors to Thaksin – able to effect a challenge.

The September 2006 coup d'état further clouded the relationship between democracy and space for advocacy. Many in an increasingly cowed NGO movement under Thaksin welcomed the coup as a way out of an impasse, even though the immediate martial law regulations prohibited gatherings of more than five persons in public places and clearly circumscribed street advocacy. There were also significant civil society voices, for example the director of the Thai Labour Campaign and Chulalongkorn activist academic Gi Ungpakorn and US-based Thai historian Thongchai Winichakul, who spoke out against the military intervention and against NGO and other civil society leaders who accommodated and even welcomed the change. Further complicating the situation was the fact that the establishment of the National Legislative Assembly, in effect a military-appointed parliament, gave places to prominent academics and NGO leaders such as Surichai Wun'geo and Tuenjai Deetes on the top law-making body.

Advocacy about development and on wider policy issues has thus emerged in Thailand largely, but not exclusively, through NGOs. The advocacy role of NGOs must also be understood in relation to wider civil society currents, which are in turn associated in part with Cold War and post-Cold War political developments and oppositional currents. In particular, Thailand's development advocacy has drawn on the twin social and environmental impacts of that country's development direction. Key points of challenge have included dams and forestry issues, both of which span the socio-environmental divide, but both of which also have been a cause for division within the NGO movement.

In many important ways, there is a logic to the spilling of advocacy across borders in a regional sense, as the impacts of Thailand's development have similarly extended into the wider Mekong Region (Hirsch 2001). Thailand's rapid growth in demand for energy and natural resources expanded the country's resource economy beyond its national borders, into the neighbouring countries of Burma, Laos and Cambodia, and to a lesser extent, into Vietnam and southern

Figure 8.3 'Heroes of the Poor' at the Thai Ban Local Knowledge Centre in Ubonratchathani Province

China. Thailand's successful environmental movement also spread beyond its borders. The political rapprochement with neighbouring countries following the end of the Cold War at the end of the 1980s facilitated this shift. At the same time, major development programmes such as the Greater Mekong Subregion were getting under way. There were obvious concerns that Thailand's fast-track growth experience was going to be repeated in neighbouring countries, and both Thai and international NGOs recognised a policy advocacy agenda around wider Mekong development. However, the political spaces beyond Thailand's national borders set very different contexts for NGO advocacy.

Advocacy by other means? Influence in closed political spaces

In Thailand, one NGO was quick to recognise the mixed impact of its own and others' advocacy success in curtailing dam construction and revoking logging concessions within Thailand's borders. The Project for Ecological Recovery established a sister organisation, Towards Ecological Recovery and Regional Alliance (TERRA). TERRA's main role was to address the impacts of Thailand's sourcing timber from Burma, Laos and Cambodia, the vision of Laos as a hydropower-based energy supplier to Thailand and, later, to other neighbouring countries, and a more general regionalisation of Thailand's resource economy in fisheries, gem mining, tourism and other areas.

TERRA and others quickly realised that policy advocacy in Burma, Laos, Vietnam, China and Cambodia was very different from that in the open society of Thailand. Moreover, for a Thai NGO to coordinate such advocacy creates its own problems, given the history of mistrust and difference of styles between political cultures of the region.

Regional development institutions rather than governments of other Mekong countries found themselves the early targets of a widened Mekong regional advocacy. There were three main reasons for this. First, the rapprochement at the end of the Cold War took on a regional form which very quickly resurrected development plans conceived in the 1960s, and two key institutions – the Interim Mekong Committee and the Asian Development Bank – took up the mantle to try to realise and further elaborate these plans. Second, partnership between regional advocacy NGOs and international partners in Australia, Europe and North America meant that the latter found it easier and more politically appropriate to target institutions supported by their own governments. Third, there were very limited avenues for influence of most governments of the Mekong Region.

Early advocacy focused on the revitalised Mekong Committee, as it dusted off plans to dam the Mekong mainstream. The plans included a cascade of dams that would have left little water flowing along the entire length of the Lower Mekong River below Chiang Khong in northern Thailand. By the early 1990s, megaprojects such as the Pa Mong Dam had come back on to the agenda and were being actively supported by the Mekong Secretariat, which was still located in Bangkok. Thai NGOs' experience with dams and the confidence of the movement that had successfully stopped the construction of Nam Choan helped to challenge the revived hydropower agenda. Initially the scheme was scaled down to a proposed series of 'run of river' dams in place of large impoundments, but soon the mainstream dams receded from the agenda altogether.

The regional advocacy target shifted through the 1990s, in part as the mainstream dams appeared to recede from the policy agenda because of the reconstituted Mekong River Commission, and in part as ADB's GMS programme became the institutional centre of gravity in promoting large-scale infrastructure development of concern to environmental NGOs. By the late 1990s, specific projects (such as the Theun-Hinboun Dam) and the GMS programme as a whole had attracted criticism, culminating in large-scale protests against the ADB at its annual meeting in Chiang Mai in 2000. However, the greatest advocacy efforts were concentrated on certain megaprojects, notably the World Bank-supported Nam Theun II Dam in Laos (Hirsch 2002).

National-level advocacy throughout the Mekong Region has proven much more problematic. Involvement of Thai or other foreign NGOs in advocacy against government policy is easily portrayed as infringement of sovereignty, the more so when it comes from countries that have themselves already developed economically in part by exploiting their own natural resources. Meanwhile, the experience of NGOs in the formerly centrally planned socialist countries was very limited and not seen as a legitimate part of the political process. The chaos of the former Soviet Union and Eastern Europe was seen by governments as a warning against pluralism and unregulated challenge to authority. However, each country has dealt with NGO emergence in its own way.

One-party-governed Laos proscribes the establishment of Lao NGOs. At the same time, numerous international NGOs have long operated in that country, one of the world's poorest in terms of GDP per capita and according to basic health and literacy indicators. Uniquely among the three countries of Indo-China, Laos has received NGO support throughout – albeit at quite a low level during the post-1975 period as the country isolated itself from Western influence. During the 1980s, most NGOs were involved in relief work and rural development programmes. By the mid-1990s, when the dam-building agenda appeared to be accelerating rapidly, many international NGO workers based in Vientiane became increasingly involved in attempting to influence the policy agenda by providing a counterweight of information about the negative experience of dams in Thailand and other countries. Individuals associated with established programmes such as CUSO and the Quakers played an important leadership role in this orientation toward the policy agenda.[1] At the same time, Laos's reliance on external funding meant that there were points of leverage beyond the Lao government. A number of Lao nationals working for international NGOs, including some organisations which had successfully completely nationalised their in-country staff, joined discussions around dam issues but also in connection with other projects of concern – such as the World Bank's ill-fated Forestry Management and Conservation Project (FOMACOP).

Experimental forms of non-confrontational organisation and mobilisation were established, notably the Sustainable Agriculture Forum, which had no formal institutional identity but which provided a strategic focus for NGOs to seek spaces for policy influence. Further, dialogues were established with the support of Thai NGOs. The Community Forestry Support Unit established in the Department of Forestry worked closely with Thai NGO personnel. In February 1996, TERRA combined with the Committee for Planning and Cooperation, Laos's national planning and investment agency, to hold a week-long seminar and training programme on 'Planning for Sustainable Development'. The aim was to expose central and provincial-level Lao planners to alternative voices and perspectives.

Gradually, however, NGO advocacy in Laos has been circumscribed. Never really based on leadership or primary initiative on the part of Lao nationals, the advocacy became increasingly painted as foreign interference. This was sharpened greatly by the acrimony generated by international NGO mobilisation against the Nam Theun II Dam. The World Bank Board's approval of the dam in May 2005 followed more than a decade of sometimes acrimonious and confused debate between various parties, which involved splits within the NGO movement and illustrated the difficulties and dilemmas of international NGO mobilisation 'on behalf of' marginal groups in a closed country such as Laos (Hirsch 2002). The campaign ultimately ended with approval of the dam being seen as a kind of green light to go ahead with several other hydropower projects in Laos. It also marked the World Bank's return to the fray of supporting large dams, after an absence of more than a decade during which the World Commission on Dams report had been provided. Nevertheless, Nam Theun II can also be seen as a case of partial advocacy success in that the safeguard policies, the international spotlight on the project and the considerable resources committed to mitigation and compensation are in fact an outcome of the pressure put on the World Bank, the project developers and the Lao government by NGOs inside and outside the country.

Unlike Laos, Vietnam does allow indigenous NGOs. However, these are very different types of organisations compared to the NGOs in Thailand. To understand NGO advocacy and its limits in Vietnam, it is important to recognise the Cold War roots of their presence. It is also important to recognise that non-state action and advocacy take forms quite specific to Vietnam and are certainly not limited to NGOs (Kerkvliet *et al.* 2003).

Following Vietnam's occupation of Cambodia in 1979, bilateral assistance to both countries was severely curtailed, despite enormous humanitarian and reconstruction needs after decades of wartime destruction in both countries. A few international NGOs helped fill the void, including the Oxfams and peace-oriented groups such as the Quakers and Mennonites, which had opposed the United States' role in the war. As a result, progressive international NGOs formed a close and constructive relationship with the governments of both these countries. However, following the Paris Peace Accords of 1991 and the welcoming of both Vietnam and Cambodia back into the international community, along with Vietnam's outward-looking reform programme after 1986, bilateral agencies and other international NGOs established large aid programmes in both countries. In Vietnam, local NGOs were formed, but these were mostly associations established by retired academics or government officers in fields such as environmental conservation. There are notable exceptions, for example the NGO Toward Ethnic Women (TEW), an advocacy-oriented organisation whose founder used good Party connections and an association with the official Institute of Ethnology to establish a strong campaigning role on behalf of women from ethnic minority groups in the country's highlands (Gray 2003). The strong relationship between international NGOs and government was maintained, but it was not based directly on advocacy except to the extent to which individuals had the ear of high-ranking officials. Australia's ambassador in the early 1990s had an NGO background with the Freedom from Hunger Campaign, which helped shape a progressive government-to-government humanitarian aid programme.

However, the commitment of Vietnam to a fast-track growth strategy and reliance on hydropower as a fundament of the country's energy policy, combined with a highly circumscribed political space, means tensions have emerged between international NGOs and the government. NGOs tend to effect change through cooperation with official bodies such as the Vietnamese Women's Union, which is an official organ despite its definition as a non-governmental mass organisation. There is little by way of direct policy advocacy, and any independent moves or challenges are quickly suppressed. Most challenges to date have taken the form of spontaneous action at the local level by villagers incensed by corruption and abuse of authority by local officials, rather than by outside organisations.

In a new departure, the Vietnamese Union of Scientific and Technology Associations (VUSTA), an official body, has joined with the International Rivers Network to carry out a study of the giant Son La Dam in north-western Vietnam (Nguyen Manh Cuong *et al.* 2006). The critical nature of the report shows an interesting use of science to advocate in a hitherto highly sensitive area – resettlement of people displaced by large-scale development projects. This reflects the openings created in a rapidly emerging civil society landscape and also shows the differentiated nature of the state as well as NGOs, even in a country often perceived as having a monolithic polity.

Like Laos and Vietnam, China is a one-party state that entertains no political pluralism at the political party level. However, the severity of China's industrial pollution and resource degradation problems has spawned an environmental movement, including the emergence of a number of quite outspoken NGOs (Yang 2005). These organisations are mainly involved in environmental restoration work, but have increasingly campaigned at the policy level. For example, a series of dams planned for the Nu Jiang (Upper Salween River) was put on hold, in part as a result of lobbying by Yunnan-based NGOs.

A key to the ability of environmental NGOs to lobby in China has been their cooperation with different levels of government. At the central level, this has included an increasingly constructive relationship with the State Environmental Protection Authority. Cooperation and strategic alliances also extend to the level of prefecture government, where local officials are concerned at having to bear the social, economic and political costs of projects that have been pushed at the provincial level and that lead to grievances by villagers. Some of these NGOs, for example, Green Watershed have good international connections in the Mekong Region but international linkage – and in this case an international reputation that came with the award of the Goldman Environmental Prize in 2006 (Goldman Environmental Prize 2006) – can be a mixed blessing, and such NGOs have been circumscribed when they have pushed beyond politically acceptable limits.

Given that NGOs are significantly circumscribed in these three countries, can other areas of civil society become a source of advocacy on the part of the environment, the marginalised and the rural poor? To an extent, universities also provide independent, even critical, voices in Vietnam and China, but less so in Laos. However, there are clear limits to the areas in which critique is entertained, and this is one reason that the environment has continued to provide a space for action where more socially and politically charged areas of critique remain no-go zones. Local government has also resisted or reformulated certain aspects of central government policy in each country, and in Vietnam and China significant incidents of village-based resistance are reported in the media. Of course, the presence of foreign-based NGOs that employ local staff also affords a degree of surrogate NGO advocacy within all three of these countries, but under the constraints outlined above.

Cambodia has seen the emergence of a more vibrant NGO community, within a multi-party system. A number of local NGOs have emerged in tandem with the large foreign NGO community, and lobbying is vibrant in the area of natural resources policy – particularly around forestry and fisheries issues. As a coordinating body, NGO-Forum Cambodia has developed lobbying skills. A number of NGOs were established specifically to lobby. For example, the Sesan Protection Network deals with problems in north-eastern Cambodia faced by indigenous minorities affected by hydropower development in Vietnam's central highlands (Hirsch and Wyatt 2004). The Fisheries Action Coalition Team (FACT), an NGO based around the Tonle Sap (Great Lake), both carries out local community fisheries livelihood development and campaigns about threats to the fishery and unjust access problems resulting from the fishing lots system.

Burma (Myanmar) has no independent NGO system. However, numerous Burmese NGOs base themselves in Thailand and campaign on human rights, ethnic minority grievances and environmental issues in Burma, but with little discernible effect in Burma.

Toward a Mekong civil society advocacy?

To date, regionalism in the Mekong has been highly unequal. There is more intermingling and cross-border cooperation among elites than is possible among subaltern voices. To what extent is a regional Mekong civil society emerging through linkages between NGOs in their very different political spaces in the Mekong Region, and to what extent is advocacy the basis of such an emergence? Thai and international NGOs, foundations and academic centres and networks have led the way in establishing Mekong-wide networks. Given the chequered history of Thailand's relations with many of its neighbours, Thai leadership of regionalised advocacy is problematic. On the one hand, international NGOs from outside the region are seen more readily as 'honest brokers', yet on the other they are less engaged than those based in the riparian countries. However, home-grown NGO advocacy in the Mekong countries other than Thailand is constrained by the different political contexts in which they operate.

In searching for, envisioning and supporting a regional advocacy from within, it is necessary to look beyond established civil society forms. It is likely that a truly home-grown 'Mekong civil society' will extend well beyond the NGO sector and will incorporate universities, think-tanks, people's organisations, chambers of commerce and a range of both local state and non-state actors pressing for change from below.

There are still many obstacles to an NGO advocacy movement that is able to engage at a regional level, including differentiated political spaces, difficulties in accommodating different advocacy 'styles', a strong developmentalist ideology to contend with, poor articulation of regional agencies with civil society, and ever-present threats and fears of co-optation. At the same time, however, the regionalising logic of integrative development in the Mekong is a ripe ground for a regionally as well as nationally targeted advocacy response from within.

9 Asian Development Bank

NGO encounters and the Theun-Hinboun Dam, Laos

Lindsay Soutar

On the morning of 6 May 2000 a crowd began gathering in the city of Chiang Mai in the north of Thailand. The crowd consisted of farmers, villagers, people's organisations, environmentalists, trade unionists, human rights activists, students and representatives of national and international non-government organisations. Shared concern about environmental preservation, management of resources and the living conditions of the many subject to poverty in the developing world catalysed the group's formation.

On the same morning, inside the Westin Riverside Hotel, also in Chiang Mai, another group began forming. This group consisted of bankers, economists, engineers, social development specialists and politicians. They had also gathered to discuss issues of development, poverty reduction and social, economic and environmental wellbeing. The event was the annual meeting of the Asian Development Bank. Despite the common concern of these two groups, major discrepancies existed in their perspectives about how to achieve development. These discrepancies existed to the point that the former group, the 'People's Forum', was accusing the latter group, member government representatives and staff of the Asian Development Bank, of doing more harm than good in the development process. These accusations were being voiced in a large-scale protest in front of the hotel.

Debate went on within both groups as to how to deal with one another. For the Bank there were considerations about how to respond to the protestors outside, and whether or not the Bank president should go out and 'meet the people' (Tadem 2003). For the People's Forum and their associates, debate centred on how to lobby the Bank; while some were advocating participation in Bank-facilitated events for dialogue, others saw this process as futile, preferring to remain outside (ibid.).

Debates over how to engage, and the seemingly unbridgeable chasm in the perspectives of the two different sets of actors, appear to be a permanent feature of contemporary international development discourse. Encounters between mainstream and alternative development advocates occur more frequently and with more intensity than ever before. Non-government organisations (NGOs) have assumed a significant role in challenging current development practice and articulating alternative development agendas. In the Mekong Region, where fast-tracked, growth-led development has brought many changes, NGOs have been a crucial alternative voice to mainstream processes.

This chapter explores how NGOs advocating alternative development in the Mekong Region have pursued their agenda. In particular it discusses campaigns against the Asian Development Bank (ADB), the most dominant and influential of the multilaterals in the region. In order to gain a deeper understanding of the effectiveness of NGO advocacy work, it examines Bank changes in response to NGO lobbying, the changing nature of relationships between the two sets of actors and the direct outcomes of campaigning in the region.

Background to ADB–NGO encounters in the Mekong Region

Rapid economic, social and ecological change in the Mekong Region has prompted debate over what constitutes a desirable development future for the region and the means to attain it. The region is characterised by uneven levels of economic development, industrialisation and marketisation, within and between countries, and the natural resources of the region provide for those who live in the Mekong basin and beyond in both subsistence and income terms. Over the past century, appropriation of resources has increased wealth and improved living standards for some in the region (Rigg 1997). However, economic and environmental changes as a result of resource extraction have brought about negative effects on livelihoods for others, notably the rural poor (Parnwell and Bryant 1996; Dixson and Smith 1997). These changes have not been welcomed by all. They have been condemned outright by some actors who call for a 'back to basics' or 'post-development' future for the region (Phongphit 1986) and criticised by those who point out that it is often the most marginalised who are left behind in pursuing growth and development. A wide range of civil society actors active in the region are amongst those contemplating past and future development options.

Civil society in the region is characterised by a dispersed and uneven nature between countries (Hirsch 2001). While local and national movements have been present in Thailand for some decades, and Cambodia has recently witnessed a budding local non-governmental sector, in Laos and Vietnam there has been little replication of these changes. To some extent, civil society activity in Thailand and Cambodia has spilt over into their neighbouring countries; the Thai-based regional NGO Towards Ecological Recovery and Regional Alliance (TERRA) with its publication *Watershed*, which focuses on environment, development and community issues in the Mekong Region, is one such example. To a significant extent, however, international NGOs still dominate the regional-level advocacy scene in most of the Mekong countries. These organisations, where possible, try to align with local NGOs to challenge regional development plans.

The current regional development agenda revolves around a programme for regionalisation, the Greater Mekong Subregion (GMS) Program. The Asian Development Bank has been a key player in instigating this programme, which incorporates the countries of mainland South-East Asia as well as Myanmar and Yunnan Province of the People's Republic of China. The material aspects of the programme have a strong focus on the development of infrastructure, with plans to link the region through interconnected networks of roads and rail, energy and telecommunications (Pante 1997; ADB 2002). These 'economic corridors' are intended to utilise the resource base of the region, maximise market access for the people of each country and promote the flow of goods and people in the region. The discursive aspects of the programme play on the notion that regionalisation and closer economic ties can bring 'peace dividends' to the region, as it transforms itself from a 'battlefield' to a 'marketplace' (Murray 1994).

As with past regional development agendas, large dams on the waterways of the Mekong watershed have been a particular focus for development planners (Bakker 1999). Renewed initiatives for development have included both the construction of dams on Mekong tributaries, primarily in Cambodia and Laos, and the identification of, and planning for, a number more. Large-scale infrastructure projects of this type, however, have for some time been the target of worldwide NGO and social movement discontent and in-depth advocacy work. In the early and mid-twentieth century, large dam construction was often touted as a marker of national development but, through the latter part of the century, the negative environmental and social effects were widely documented (Goldsmith and Hildyard 1984; McCully 1996). Dam sites emerged as symbolic places of struggle and resistance to those who called for alternatives to large-scale development. High-profile cases, such as the Narmada Dam in India and the Arun Dam in Nepal, have caused public relations disasters for international lenders. Mounting concerns over the role of large dams in the development process culminated in the late 1990s in the establishment of the World Commission on Dams (WCD). This multi-stakeholder forum made a number of findings about past dam projects and their effects and also produced recommendations to guide future large dam development.

The market-led development model focusing on large infrastructure development adopted by the Asian Development Bank through the Greater Mekong Subregion Program is typical of the approach adopted by multilateral development institutions globally. In many respects, including structure, function and objectives, the ADB has followed the mould established by the Bretton Woods institutions, particularly the World Bank. Like the World Bank, the ADB holds significant economic and political sway over borrowing countries. Also as with the World Bank, its activities and lending have been subject to significant civil society scrutiny and been the target of sustained NGO advocacy efforts.

The negative impacts of the mainstream development agenda promoted by multilateral development banks have been well documented (Danaher 1994; George and Sabelli 1994; Rich 1994) and include environmental degradation, social dislocation and economic marginalisation. Critics have argued that the multilateral banks are unaccountable and undemocratic and reject the banks' growth-centred approach to development. It has also been argued that the highly bureaucratic nature of the banks and their concern with protecting their own position and power have resulted in institutional ineffectiveness.

Recently the development banks have undergone a number of changes, some in response directly or indirectly to NGO advocacy work. Research on NGO campaigns against World Bank policy and projects (Fox and Brown 1998; Royo 1998; Rumansara 1998; Nelson 2000b; O'Brien *et al.* 2000; Tussie and Casaburi 2000; Khagram 2004) has found that external pressures on the Bank have led to it suffering an 'identity crisis', forcing it to re-examine its roles and make changes with regard to its relationship with civil society and borrowing and donor countries. The most significant changes (which also apply to the Asian Development Bank) include an opening up to civil society, review and establishment of policies, new accountability and transparency mechanisms and a new mandate of 'poverty reduction'. The focus and attention of various civil society actors through multiple campaigns have had a small but significant effect in pressuring the Bank to become more accountable (Fox and Brown 1998). Nonetheless, however, theory about large and bureaucratic institutions suggests that, overall, these types of organisations are resistant to change, and change that does occur more often represents organisational adaptation than a meaningful alteration in direction (Barnett and Finnemore 1999).

The following discussion is based on the results of research investigating the changing features of the relationship between NGOs and the Asian Development Bank in the Mekong Region. The research used qualitative methods to look within both NGO networks and the Bank and understand interaction at two scales. The first scale was at the institutional level at which Bank and NGO perspectives on interaction were explored. The second scale was at the project level in which a case study of interactions between the Bank and NGOs concerning a specific Bank-funded project was detailed. The Theun-Hinboun Hydropower Project, a dam development in Laos, was selected as the case study on account of the protracted and substantive nature of the interactions between the Bank, NGOs and other actors following the dam's completion. Research data was collected through in-depth interviews with 16 NGO staff from a range of organisations campaigning in the region, with over 18 ADB staff from different sections within the Bank and with a number of other stake-holders involved in the dam development project. Data was also derived from written sources including Bank and NGO correspondence, particularly for the purposes of the case study. Field work was carried out over six weeks in mid-2004 in the capital cities of three countries: Bangkok in Thailand; Manila in the Philippines; and Vientiane in Laos. The field locations selected correspond with the locations of key NGO offices, the ADB headquarters and Laos where the Theun-Hinboun is located.

ADB–NGO encounters and perspectives on an evolving relationship

Interactions between the ADB and NGOs occur through a number of forums and at different scales. In the Mekong Region, NGO campaigns have been directed at selected projects of the Greater Mekong Subregion Program, particularly those involved with hydropower and transport development. Attention has increasingly been paid to the Bank's activities in Laos and Vietnam where, in the past, owing to the weak civil society in these countries, the Bank has had something of an 'easier run' in administering its projects. In Thailand, however, this has been far from the case, and direct challenge and open confrontation have occurred, both over particular projects and at the afore-mentioned protests at the Bank annual meeting in Chiang Mai, Thailand. Like other international campaigns against the multilateral development banks, NGO campaigns against the ADB in the Mekong have adopted a two-pronged approach. Project-level campaigns have aimed to both shed light on and call for resolution of problems arising from individual projects funded by the Bank. Institutional-level campaigns have targeted the Bank itself, calling upon it to reform policy, procedure and practice.

While interaction and conflict between the Bank and NGOs are at their most visible at public events such as protests, relationships are also carried out in different ways behind the scenes. NGOs lobby the Bank by making direct representations to Bank staff and executive directors, and the Bank has opened up formal spaces for access through hosting policy and project consultation sessions and establishing the NGO Centre. Opportunities for NGOs to engage in contract work on Bank projects have also been established. Some interactions are ongoing processes, while others are place- and time-specific; some are driven primarily by the Bank, some by NGOs and some by both. Some interactions are characterised by high conflict and others by high cooperation. Each interaction can be viewed differently by the different actors involved.

An exploration of the perspectives of NGO actors reveals the thinking behind decisions on where, how and when to lobby the Bank. What do organisations hope to achieve by their lobbying and how do they want to get there? While civil society is at times referred to as a homogeneous and united entity, in reality it is composed of diverse, multi-faceted and, at times, conflicting interests (McIlwaine 1998). Within NGO networks there have been divided views as to how to target the Bank and debate over whether campaigns should aim to reform the Bank or to see it abolished. Those who wish to see it abolished consider that communities and developing countries would be better off if the Bank could be done away with; as one NGO staff member commented: 'my inclination is to disempower the Banks and transform them into quite different types of agencies . . . tantamount to closing them down'. Other NGOs, however, take a more pragmatic view and consider that, as an institution embedded in current political and economic structures, the ADB will not easily be dismantled. As one NGO staff member suggested: 'We cannot kill them. . . . They are already set up. . . .

Figure 9.1 Protestors and police at the Chiang Mai People's Forum in 2000

They are just a tool of neo-colonialism. Take away one tool and they'll build another.' For these actors, the goal of campaigns then becomes trying to pressure the Bank to reform, advocating alternative development approaches and 'keeping [the Bank] in a direction that is not too harmful to too many people'.

Trying to pressure the Bank in this way has also created division, however, as different organisations have debated the best means to pressure the Bank to reform. The most extreme position, held by the more radical organisations, is not to engage with the Bank at all, but to maintain critical resistance and highly confrontational relationships, as reflected in the following statement:

> For me and for my organisation, our biggest lesson is that we should not lobby the ADB. The ADB is sitting in an undeserved position of policy power by virtue of its financial resources. To lobby them indicates that we accept their position of power, which we don't.

Others will engage in Bank consultations, depending on the circumstances and perceived potential outcome of engaging. For both these groups, fear of being co-opted is significant. They are wary that interaction with Bank staff may be used to legitimise Bank practice and that the issues being raised by NGOs will be ignored. As one staff member put it, the organisations 'are conscious of not wanting to be pawns in [the Bank's] public relations campaigns'. They maintain a fiercely critical independence from the Bank and consider that 'the only way to deal with the Bank is through power – things like public opinion and justification, but not negotiation. We do need to talk with them, but talk with them not as friends but in opposition.'

This approach diverges from that adopted by more moderate NGOs which believe reform is best pursued through working with institutions and attained through access to decision makers and building trustful relationships. The rationale for establishing working relationships with the Bank was explained by a staff member of one of the more moderate groups:

> By having a dialogue we hope that in the future we might be able to influence their policy. The way we work we want to be constructive but reserve the right to criticise the Bank. But I think before we get there we have to exhaust all forms of dialogue. . . .

This position has, however, been highly contentious and has given rise to a degree of tension between NGOs. Accusations of 'sleeping with the enemy' reflect criticism of their position, with some NGOs suggesting that those organisations that form closer relationships with the Bank weaken the overall advocacy efforts of NGO campaigns. Thus, 'if we could all work in harmony, for example by all boycotting [a consultation], it would give [the Bank] a stronger message. There is a danger that NGOs become rubber stamps, giving legitimacy to the Bank.'

The more moderate NGOs have defended their position, arguing that different groups dealing with the Bank in different ways can actually strengthen overall advocacy efforts:

I don't think our efforts really undermine the efforts of others. . . . We try to complement them. . . . By having an organisation that is able to bring points to the ADB in a less threatening manner it strengthens the advocacy of [other] groups.

Criticism from some corners also extended to NGOs that competed for Bank contracts. Again, the issue of contention was the danger that nominally independent organisations would be co-opted and lose their potential to advocate effectively. As one NGO staff member pointed out, vying with the private sector for Bank money meant that, 'for [these organisations], there is definitely an interest not to be critical'.

This issue of NGO contracts with the Bank is particularly pertinent when examining Bank perspectives on NGO involvement. How have Bank staff understood and responded to the range of NGO positions outlined above? For some years now, the Bank has framed its evolving relationship with NGOs in a positive light. Policy documents such as the 'Policy on cooperation with NGOs' (ADB 1998a) and the 'Framework for cooperation' (ADB 2003a) indicate that the Bank is seeking to pursue

an expanded program of cooperation with NGOs . . . with a view to strengthening the effectiveness, sustainability and quality of development services ADB provides. The objective of ADB's cooperation with NGOs [is] . . . to work with NGOs to incorporate NGO experience, knowledge and expertise into ADB operations.

(ADB 1998a: 4)

The ADB is a large and bureaucratic institution, however, with staff from many and varied backgrounds. Staff perspectives are structured by their positions within the Bank, their views on development and their experience in dealing with NGOs, and not all staff view NGO activity as favourably as the formal policy would suggest. Interviews with staff from a range of departments indicate that internal debates are occurring within the Bank about how they should relate to NGOs. A general, but not all-encompassing, trend is that staff from the policy section of the Bank tend to view NGO activity in a more positive light than do staff from operations sections of the Bank. Two key internal debates revolve around the issue of 'functional' and 'dysfunctional' relationships and around issues of NGO accountability.

For most staff, the most functional relationships the Bank has are with NGOs whose services are engaged under formal contract to implement and/or monitor projects. These 'operational' NGOs are seen as offering on-the-ground technical skills, knowledge and experience. Bank staff value the contribution of these organisations as they see them as 'efficient and cost-effective' and 'at the grass roots and able to reach those who ADB cannot'.

These views contrast with those relating to advocacy NGOs, which frequently challenge the Bank and are seen by some ADB staff members as difficult, critical and non-constructive. Some staff members argued that NGOs have a tendency to generalise, run a one-point agenda and sometimes be blind to 'big issues'. The different perceptions of operational and advocacy NGOs are exemplified by a quote of one operations department staff member:

> Working NGOs have real work experience and are usually engaged in concrete dialogue with the ADB, whilst advocacy NGOs usually advocate views which are abstract, academic or theoretical and hence have limited operational applicability and are difficult to relate to. With working NGOs we share common experiences and learn from each other.

Some staff also suggested NGOs could be unrealistic in their expectations of the Bank and that the idealism of NGOs prevented them from making meaningful contributions to the development process. As one commented:

> I know a bit about the ideal and the possible and sometimes someone will come in and give us the NGO textbook for an ideal world and you think 'Well, if I ruled the world then I'd like it like that too' – but then there is reality. . . .

Advocacy NGOs were looked on more favourably when they were engaged in 'functional relationships', that is, when they were not overly critical, provided constructive suggestions or brought up issues that the Bank might have over-looked. The difference between dysfunctional and functional relationships is illustrated by the following two quotes: '. . . dysfunction is where [NGOs] go into the press and say the ADB is the evil empire of rape and pillage of poor people in communities' and

> . . . [function is] when they keep their emotions in check, not being overly dramatic, and when they get their facts straight. I appreciate that activists can be quite passionate about the issues they work on, and have good grounds for the concerns they raise. They can sometimes go overboard in the way that such concerns are expressed.

Internal debates also exist within the Bank in relation to issues of account-ability. This is important, as the extent to which Bank staff consider NGOs to be important stakeholders in the development process will influence the extent to which their concerns or interests are taken on board. Some staff, often those who work closely with borrowing governments on project implementation, questioned the right of NGOs to express views, particularly critical ones, about Bank activities in borrowing countries, as illustrated by this quote:

Say you have project 'X', and the government feels that it will really help the people but then an NGO in Germany causes the project to be derailed or calls for making it more effective. Who will pay the cost of this? Ultimately it comes to the government . . . so one thing the NGO community lacks is who they are accountable to.

This contrasted with other opinions, again predominantly of those working in policy areas, who suggested NGOs could raise real concerns about development processes and both had the right to do so and played a useful role:

As important as accountability is, the message [what an NGO says and what it does] needs to be held on an equal plane. Who you talk for is important. Equally important is what you are saying and what relevance it has to the issues.

Many staff within the Bank made clear that, for the Bank, interaction with NGOs and actors at the local level is secondary to interaction with governments, which are the Bank's key partners. As one staff member put it, '. . . We are accountable to member governments, so working with NGOs is subordinate to working with governments, who are our major clients.'

These views on NGO accountability are revealing. While the Bank officially affirms NGOs as important stakeholders in the development process, within some sections of the Bank views prevail that the Bank and governments are the most legitimate and significant stakeholders. For these staff there appears to be little recognition of the challenge of managing *different* stakeholder interests. As mentioned, scepticism about NGO legitimacy is often the attitude of staff who work most closely with borrowing governments. Borrowing governments also hold mixed feelings about working with NGOs and the Bank's role in incorporating them into the development process, as indicated by one borrowing-country executive director:

The influence of donor governments has gone a long, long way in fostering an NGO culture in the MDBs [multilateral development banks]. This is not always helpful – by and large it is – but it is not always favoured in these [borrowing] countries. In many places [the Banks] are forcing the government to swallow this NGO culture.

In many ways tensions over new avenues for interaction and evolving relationships are exemplified by debates about the ADB NGO Centre. The NGO Centre was established in 2001 to facilitate interaction and to foster constructive relationships between NGOs and Bank staff; however, the role, purpose and value of the Centre have been understood very differently by those involved.

For most Bank staff, the establishment of the NGO Centre is indicative of the Bank seriously attempting to be responsive to external actors. As one staff member suggested:

The ADB is able to work more effectively with the NGOs now than before given its recent reorganisation. There is now a clear line of communication between NGOs and ADB, with the NGO Centre as the focal point which enables NGOs to discuss issues more frequently and have direct access for more frequent dialogues.

Not all NGOs share this view, however. Some organisations instead view its establishment as a technical move by the Bank to be seen to be accommodating NGO interests but in reality trying to 'manage' NGO advocacy and act as a buffer or 'firewall', limiting access to influential Bank staff. The organisations that hold this view will not work with the Centre. As one NGO activist commented, 'We try not to go through the Centre. We don't like the idea that there is only one part of the Bank to deal with NGOs – the whole staff should be accountable.'

Staff of the NGO Centre rejected the suggestion that the Centre should be seen as a 'firewall' and instead tried to reinforce its value as a contact point for those less savvy about the Bank's operations:

The savvy NGOs can speak with whomever they please, as they have the contacts, knowledge and confidence. But being a complex institution, we need to ensure we are being responsive. . . . Maybe what we call coordination they call obstruction, but we have nothing to gain by keeping people away from the decision maker.

While some NGOs were reluctant to work directly with the Centre, others thought the Centre could play a useful role and would work with it when they saw it as advantageous. Most NGOs, however, were concerned that the capacity of the Centre was undermined by its small size, lack of resources and limited influence within the Bank. As one NGO engaged in more cooperative relationships with the Bank indicated:

The staff from the NGO Centre all [have] an NGO background and I feel this helps with credibility of the NGO Centre. However, what is noticeable is that there is very little power for the Centre. . . . I think they are doing as much as they can with the resources available but that doesn't mean that the rest of the Bank is necessarily following suit.

The establishment of the NGO Centre can thus be seen as a further attempt by the Bank to engender more constructive relationships with NGOs, ironically a move perhaps made in response to NGO calls for greater openness. While some NGOs have responded positively to this, for others it reinforces the divide between the Bank and external actors.

The range of views held by both sets of actors demonstrates an ongoing realignment in Bank–NGO relationships. Some interactions are characterised by cooperation and others by high conflict. So what impact have these changes had in terms of development outcomes? To gain a deeper understanding of the effect of NGO advocacy approaches, the relationship is now considered in the context of an NGO campaign directed at an ADB-funded dam development in Laos.

Development model or misguided development? The Theun-Hinboun Hydropower Project

The case of the Theun-Hinboun Hydropower Project provides a good example of the possibilities and limitations of carrying out advocacy work in developing countries, in particular of how international and regional NGOs can operate in a country with limited political space for local group activity. More generally, the Theun-Hinboun case reflects the contested nature of development and the ADB's role as a development lender and adviser. While the ADB lauded this high-profile dam as a model project, a number of NGOs were highly critical of the dam and the processes associated with its construction. In the late 1990s and early 2000s, the Theun-Hinboun Dam became the target of a sustained NGO campaign questioning the rationale for hydropower development in the Mekong and the Bank's approach to development.

Without ADB involvement, it is unlikely that the Theun-Hinboun Hydropower Project would ever have happened. The bank played a critical role in coordinating the project and in financing the Lao government portion of the project. The project was novel in that it was the first large-scale dam project to be developed in joint venture between the Government of Laos (GoL) and the private sector as a build–own–operate–transfer venture. The Lao government state-owned electricity utility, Électricité du Laos (EdL), owns 60 per cent of the $US240 million project (its investment largely financed by a $US 60 million loan from the ADB), and the ownership of the remaining 40 per cent share is split equally between the Thai company MDX Lao and Nordic Hydropower. Theun-Hinboun Power Company (THPC) was established to run the project. The rationale behind the project was to raise foreign exchange revenue for the Lao government through the sale of electricity to Thailand. This revenue could then be used to further develop the Lao economy.

The Theun-Hinboun Hydropower Project is located between the Khammoune and Bolikhamxay provinces in Laos, 100 kilometres upstream from the confluence of the Theun and Mekong Rivers. The 210-megawatt project generates electricity by diverting the water of the Theun-Kading River into the Hai and Hinboun Rivers. Energy generated is then transmitted to Thailand though large power lines passing through Thakek. The diversion of water necessary to power turbines has caused significant alteration to the river's natural flow regime. The resultant impacts of this alteration, as well as the ADB planning and development processes, prompted the NGO campaign on the project.

The Theun-Hinboun campaign began even before construction of the dam was completed. Two years prior to the 1998 opening, the Norwegian aid-monitoring NGO the Association for International Water and Forest Studies

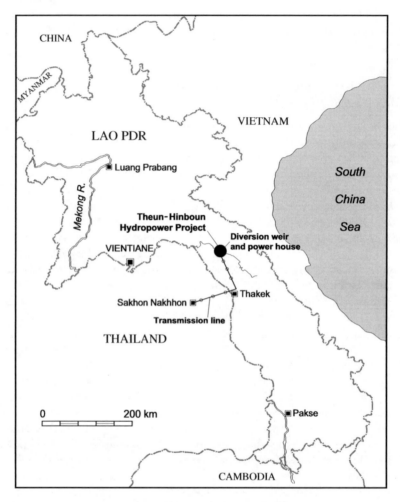

Figure 9.2 Map of Laos and the location of the Theun-Hinboun Dam

(FIVAS) released a document, *More Water, More Fish?*, studying potential impacts of the dam (FIVAS 1996). The document's title was prompted by a claim that holding back water in the dam would improve fisheries rather than cause them damage. *More Water, More Fish?* strongly rejected that suggestion, suggested impacts of the dam might be serious, and criticised the Bank's decision-making process. The Bank, however, largely ignored the report, maintaining that the project was a 'low impact, run of the river' project, not requiring resettlement and therefore not a case to be concerned about.

This position was reinforced in 1997, one year before project completion, when the Bank claimed that the project would be a 'model project with little for

Figure 9.3 Theun-Hinboun Dam wall

the environment lobby to criticise' (Gill 1997) and then, again, at the dam opening in 1998 when the Bank president declared that 'from an environmental perspective the Theun-Hinboun is ecologically friendly'. These statements raised the ire of the environment lobby, which suspected the case might be otherwise, and a full-blooded NGO campaign was launched. The campaign was headed by the International Rivers Network (IRN), a US-based advocacy organisation concerned with the protection of rivers and river systems world-wide. Its work was supported and promoted by a network of other international environmental and development NGOs, based both in and beyond the Mekong Region, including Mekong Watch, a Japanese-based aid-monitoring NGO; Oxfam Community Aid Abroad in Australia; Probe International, a Canadian-based NGO; TERRA, a Thai-based regional NGO; and World Wildlife Fund, Thailand.

The NGO critique of the project typified anti-dam campaigns internationally. Local villages in the project area are largely subsistence based, relying on the rivers and the land for their livelihood. While there was no formal resettlement in this case, dam impacts affected the ability of people to derive a living from the local environment. Impacts from the changed river flow included river bank erosion, decline in water quality and interruption to fish migration. Fisheries and fishing equipment were lost or damaged, as were riverside gardens and river access owing to bank erosion and flooding. The NGO lobby was also concerned with the overall growth-centred rationale of the project, fearing the income generated for the country would not be used for local and regional poverty reduction but rather appropriated by elite interests elsewhere.

IRN completed an in-depth critique of the project, *Trouble on the Theun-Hinboun*, after the dam began operating (Shoemaker 1998). This report, written by an independent researcher on contract, confirmed a number of concerns raised earlier by FIVAS and documented that negative impacts were rapidly becoming visible, even only a month after the gates had been closed. This report and follow-up letters were submitted to the Bank management and executive directors and called for the developers to address the issues it raised. In particular, they called for the Bank to execute compensation and mitigation measures.

In response to these protests, in May 1998, the Bank sent a loan review mission to the site. However, the initial back-to-base reports were defensive, and dismissive of NGO claims, and did not acknowledge the impacts identified in *Trouble on the Theun-Hinboun* (ADB 1998b). Furthermore, staff on this mission approached villagers who had been quoted or pictured in *Trouble on the Theun-Hinboun* and pressured them to retract statements regarding the dam's impacts. Retracted statements were then used in the Bank review mission report to refute the claims of the NGO report (ADB 1998c). Despite this initial dismissal by the Bank, NGOs maintained their lobbying, which resulted in another Bank review mission being sent some months later, in November 1998. This mission resulted in an about-turn on its previous position, the Bank admitting the validity of NGO assertions about the project's impacts. It also acknowledged the need for a comprehensive mitigation and compensation programme to be implemented for adversely affected villagers (ADB 1998d). For some time, however, these acknowledgements resulted in little concrete action.

Throughout 1999, IRN and its NGO colleagues continued to place pressure on the Bank through written correspondence demanding action. It released a follow-up research report on the dam's impacts (IRN 1999) and obtained a Bank-contracted report outlining the impacts of the dam on fisheries. This latter report, completed in June 1999, presented these impacts as even more damaging than *Trouble on the Theun-Hinboun* had indicated (Warren 1999). Despite having contracted the report, the Bank refused to release it publicly, claiming that it was not of high enough quality. It did, however, send another two review missions to the site, in November 1999 and March 2000. These mission reports, as well as letters responding to NGO appeals, attempted to shift the burden of responsibility for mitigation and compensation on to the private operator of the dam and the Lao national government. One letter suggested that, 'while the Bank endeavours to reply to your detailed queries, it will be more efficient and expeditious if you were to address these in the first instance to THPC and GoL' (ADB 1998d). NGOs rejected this line of argument, maintaining that

> while the loan has officially been 'closed' we reject any suggestion that the ADB's responsibility for ensuring that the project addresses our concerns

has ended. That would ignore the ADB's past role as the main broker, facilitator and funder of this project.

(IRN 2001)

The NGO campaign appeared to start bearing fruit when, in June 2000, a mitigation and compensation programme (MCP) was prepared by a Lao-based environmental consultant and agreed upon by the Bank and private operators. THPC released the MCP in September 2000 and it was hosted on the ADB website. The MCP established a new body responsible for environmental and social management of the project, the Environmental Management Division (EMD). The Division was entirely under the directive of THPC, with the Lao government no longer involved in direct regulation except as a majority stakeholder in the dam. While acknowledging some progress, NGOs maintained their pressure on the project, conducting a site visit and releasing a review of the new mitigation and compensation programme. The review, though critical of some aspects of the MCP, largely welcomed the Bank and THPC's apparent shift in attitude. In private, however, NGOs remained sceptical of the extent to which the MCP would be implemented and the efficacy of compensation measures.

Throughout the campaign IRN had called for independent monitoring and evaluation of the MCP (for example, see IRN 2001). By mid-2002, a new general manager had taken over the running of THPC and, with the new manager, the company set about reworking and implementing the mitigation plan. IRN soon established a dialogue with the new management, leading, in 2003, to an agreement between the two parties to bypass ADB processes and cooperatively conduct an evaluation of the project and mitigation programme. While there was some initial debate concerning the independence of this evaluation given that consultants were financed by THPC, after deliberation and consultation with involved parties consultants approved by all parties were selected. In March 2004, the review process began. Evidence from those involved in the review indicated that THPC had been making a substantial and, in some areas, fruitful effort with regard to providing mitigation to those whose livelihoods had been affected by the project. Despite all these efforts and the seemingly closer relationship, relationships between the two parties again fractured only three weeks into the review process when THPC changed the terms of the agreement, resulting in its termination. The key point of contention between IRN and THPC concerned the subject of the review process. NGOs were interested in assessing the overall dam project, with the intention of showing that, despite mitigation programmes, the impacts of a dam were ultimately irreversible. THPC were concerned with conducting an assessment of the mitigation programme only.

Throughout the campaign there was continual support from a range of other NGOs, both within and outside the region. This support was in the form of signing letters of appeal or concern directed to the Bank and executive directors (EDs), making face-to-face representations to EDs from various countries,

Figure 9.4 Altered river regime, downstream of the Theun-Hinboun Dam

publishing information in NGO publicity and resource material such as newsletters and magazines, and attending protests directed at the Bank more generally. All these events also occurred in the context of wider movements seeking reform of the Bank. These included, amongst others, the protests at annual meetings of the Bank in Chiang Mai in 2000 and ongoing campaigning to reform Bank policies.

The series of events, actions and reactions outlined above shed light on NGO strategy and outcomes as well as on Bank approaches to dealing with NGO advocacy. The case highlights a number of issues relating to potential avenues for success for NGO campaigns, to limitations of advocacy campaigns and to issues of accountability.

NGOs made the ADB the key target for their campaign against the Theun-Hinboun Project for a number of reasons:

1   The perceived potential for leverage, and thus influence over project outcomes, was greater with the Bank than with the Lao government. Donor governments based in Northern countries are theoretically answerable to their own taxpayers and could be pressured to in turn pressure the Bank to take action.
2   Had Northern NGOs targeted or criticised the Lao government, questions of legitimacy could be raised. Targeting the Bank, rather than the government, was particularly necessary in the case of Laos, where the political

freedoms to challenge project and development agendas in the country are not available.

3   The ADB had specific policies for which it supposedly could be held accountable.

Furthermore, directing the campaign at the ADB was a political choice made by the advocacy network. Theun-Hinboun could potentially have an impact wider in scope than the single project, in that it could influence the Bank's consideration of future hydropower development in the Mekong and its practices with regard to environmental and social issues. As one NGO staff member suggested:

> Theun-Hinboun was a critical test case because the ADB was billing it as a model hydro project. Politically it was very important. . . . When determining what projects to focus upon we look at whether it will have a follow-on effect for other projects.

A range of tactics was employed in an attempt to draw the attention of the general public in Northern countries, governments and the Bank to the impacts of the project. Thorough on-the-ground research was a critical component of the campaign, raising issues that were difficult for the Bank to ignore. Attempts were also made to hold the Bank accountable to internationally adopted guidelines for good practice, in particular the World Commission on Dams findings. Making suggestions based on WCD recommendations enabled the advocacy network to refute the criticism that they only protested and did not make positive suggestions.

There were, however, some limitations to the campaign stemming from the Lao national context. This campaign was unable to draw on legal mechanisms for influence, and affected citizens had no legal avenues for recourse. The political structure meant locally-based people and even Lao-based international NGOs were not able to speak out against the project. Neither could they easily provide information to those who could. In other high-profile international campaigns, NGO legitimacy has been derived through forming close partnerships between international organisations and individuals and community groups directly affected by projects. In the case of Theun-Hinboun not only was this absent, but at certain points in the campaign project the most affected people had no knowledge of the actions being undertaken on their behalf. Lobbying NGOs, in fact, 'deliberately distanced ourselves from the communities which, while not necessarily ideal, was safer in this case, as it was detrimental for [villagers] to be associated with us'.

The case of the Theun-Hinboun raised accountability questions for all actors involved in the project. Advocacy groups working on multilateral bank campaigns typically accused the Bank of maintaining decision-making processes that were non-democratic and non-responsive to those directly involved in projects. Likewise, critics of NGO activism often used the question of accountability as grounds to undermine advocacy work. Given the circumstances of this campaign, it is clear how such critics might see a double standard in that the NGOs' mode of operation parallels that which they condemn. That said, the advocacy groups involved pointed out that they had not ever claimed to represent the affected people, only that they were 'asking questions' of the Bank. As one NGO staff member claimed, 'if there was room for civil society and local groups [in Laos] then there wouldn't be a role for us, but there isn't ... so, if we don't, who will?'

There appear to be mixed feelings within the Bank as to the extent that they are accountable for project outcomes. On the one hand, letters from Bank staff indicated that the dam operators and the Lao government were primarily responsible for implementing mitigation procedures. On the other, that the Bank responded to NGO claims, in the form of site visits, reports and recommendations, as well as later taking some recognition for mitigation, indicates that they did accept a measure of responsibility for project outcomes.

The complex nature of accountability can also be witnessed in the Lao government's role in the project. While the Bank suggests that its line of accountability is to governments rather than NGOs, the non-elected socialist government in Laos is not responsive or accountable to the people of its country. Further, since the government was removed from responsibility for the ongoing monitoring of the social and environmental aspects of the project, Bank claims that its lending is legitimised by governments being the key decision maker in project development are without real basis.

In this case study an interesting tension was created as, while NGO communities have often been critical of development agendas promoting the role of the private sector, the latter's involvement appears to have been fruitful in seeing the MCP implemented. However, rather than this being a function of the private sector's greater flexibility or measure of responsibility, it appears individual agency in the form of a new general manager played a critical role. This then raises questions about the sustainability of private sector mitigation efforts should the current manager leave the company.

So what has been the outcome of all these claims and counter-claims, words spoken, reports written and actions taken? To what extent can the NGO campaign be judged as successful? The Theun-Hinboun Hydropower Project is now the largest foreign-exchange source in Lao PDR, proving successful with regard to its original goal of boosting export earning for the Lao government (ADB 2002). Much to the consternation of NGOs involved in the campaign, the 2002 Bank evaluation found the project to be 'successful to highly successful' based on the consideration that it has met the criteria of the initial project objectives (ADB 2003b). While this evaluation makes reference to the social and environmental impacts of the project, these are not considered significant enough to warrant it being considered non-successful. The Bank continued to use the project in its promotional material, and the project continues to be promoted as a model for future development, demonstrating the benefits of public–private partnerships and the involvement of multilateral actors (ADB 2002). Promotional material also states that communities are much better off since the dam was installed (ADB 2002: 24). This is a claim challenged by many NGOs, which continue to see the dam as a model of the inappropriate approach of the Bank to development. Significant questions remain as to how the project is contributing to poverty reduction, the Bank's core mandate, particularly given the significant negative impacts those within the project area have suffered.

Both Bank publications (for example, ADB 2002) and Bank staff now laud the role that NGOs played in the project follow-up process. As one staff member commented, 'this is an excellent example of how NGOs can assist governments and the ADB to ensure what they [ADB and governments] say is going to be done is done'. It appears that for public relations purposes, and to accord with its public position on 'NGO cooperation', the Bank has now embraced the importance of NGOs' involvement in the project. There is, however, something of a disjuncture between the new framing of the relationship and the reality of a case in which conflict, rather than 'cooperation', defined the interactions between the actors involved.

This case study illustrates limitations to institutional attempts to forge closer relationships with NGOs. The influence of NGOs was predominantly derived through their own initiative and made little use of Bank-facilitated channels for interaction. Furthermore, the shift within Bank policy towards participatory practices and the importance of consultation were not translated into practice in this case; local villagers had minimal input into any decision-making processes and little knowledge of the project itself.

In terms of influencing the Bank more generally, most actors involved believed that the campaign work undertaken by NGOs on the Theun-Hinboun Project had some impact on the Bank. IRN considered that

> there have been improvements in Bank practice. . . . Diligence on projects is more than it used to be. . . . They are definitely a lot more careful about what they do . . . [and] if there hadn't been campaigns and pressure on the Bank's support for hydropower in the Mekong there probably would have been more dams.

IRN also thought the campaign had made the Bank more aware that NGOs are stakeholders who can influence a project, and another NGO interviewee observed: 'Maybe it [the campaign] made the Mekong Department more afraid of NGOs. That is a good thing in itself – the Mekong Department might now be more careful where concerns of sticking with its own policies come up.' Likewise Bank staff implied that there was increased knowledge and awareness among staff that NGOs are monitoring the activities of the Bank. This has the potential to influence their practice, as indicated by one operations Bank staff member: 'It's like any activity: if there is somebody watching you, you do it differently'.

The Bank claimed that it had 'learnt lessons' from its experience with this project and that NGO involvement played a significant role in helping the Bank evaluate its practices and modify them to ensure that the types of problems associated with the Theun-Hinboun Project did not arise in future projects. While 'learning lessons' might enable a faster resolution of problems arising in specific projects, this will not necessarily mitigate larger problems created by the mainstream approach to development promoted by the Bank. In terms of influencing the Bank to reconsider its agenda with regard to infrastructure development in the Mekong Region, there is a strong indication that NGO lobbying and campaigning work has had little influence. That the Bank is now involved with the highly controversial Nam Theun II Project emphasises that the Bank (and the Lao government) still considers hydropower development to be a viable and worthwhile path to development. For NGOs, the overall irony of this case is that the Bank is now using the Theun-Hinboun Project, and the 'successful' mitigation and compensation measures implemented, to prove it has 'learnt lessons' and thus to justify its involvement in Nam Theun II.

Conclusion

So, what does an examination of both Bank and NGO staff attitudes and the case of the Theun-Hinboun reveal about the changing nature of relationships between the ADB and NGOs? Firstly, it is important to note that both sets of actors covered a wide range of individuals, groups and interests holding different views, opinions and ideas. While the relationship between the Bank and civil society often appeared highly polarised, behind the public façade of every organisation is a complex interplay of differing and competing views, agendas, opinions and approaches. Relationships and their outcomes are also contingent on the space and time at which the interactions occur.

At the same time a number of generalisations can be made about changing Bank–NGO dynamics. NGO campaigns have called for greater openness from the Bank, and the Bank has responded by placing an increased emphasis on the role of NGOs in its development practices. While this has occurred, however, the private face of the Bank has not shifted into direct alignment with its public face, some staff retaining reservations as to the value and worth of engaging NGOs as well as to their accountability. The bureaucratic nature of the ADB means that entrenched organisational views and behaviours continue to be reproduced.

From the NGO perspective, while there is a common belief that the Bank needs to undergo reform, there are many differing perspectives on the most appropriate form and direction of interaction. While some NGOs engage in new avenues for dialogue, others, wary of co-option and being caught up in Bank public relations exercises, choose to retain conflictual relationships and avoid participation in Bank-initiated forums. NGO rejection of avenues for engagement leaves some Bank staff perplexed as to how they can 'satisfy' NGOs.

Institutional perspectives have shaped project-level interactions in the Mekong. The Theun-Hinboun Hydropower Project provides an example of a conflictual relationship. In this case the ADB, pursuing a growth-led development agenda in Laos, was challenged by NGOs which adopted a variety of strategies to influence the Bank and the project outcomes. These strategies played a role in partially resolving grievances associated with the project's negative impacts. The Theun-Hinboun case indicates that campaigns conducted by networks of international NGOs, with persistence and good research, can have an impact on project outcomes. While academic research on transnational advocacy has shown that the influence of transnational networks is likely to be greater when campaigns are carried out by coalitions between international and local-level actors in democratic countries (Khagram 2004), the case of Theun-Hinboun suggests that there is potential for NGO influence in less democratic circumstances. Finally, however, it is important to note that a range of actors plays a part in project outcomes, making it difficult to generalise about scales of influence and too complex to ascribe complete responsibility for final outcomes.

In examining both institutional positions and the Theun-Hinboun case it is apparent that, while aspects of Bank–NGO relationships are characterised by cooperation and can be framed in a 'positive' light, disparate internal views within the Bank and the desire of some NGOs to maintain critical distance combine to ensure that non-cooperative relationships and divided views remain a characterising element of Bank and NGO encounters. The Bank's approach to development, although challenged by NGOs, ultimately remains unaffected and, unless the ADB or NGOs substantially change their worldview, conflict will continue to be a feature of the relationship. As the case of Theun-Hinboun shows, however, this conflict should not be seen as problematic, but rather as contributing to a 'creative tension' in an evolving relationship.

10 Making poverty history?

Barbara Rugendyke

So you think poverty is just too complicated and big? Can't see how you can possibly influence world leaders, international bankers, European trade policy when you have enough trouble influencing your mum, your boy/girlfriend, your children, your wife, yourself?

Well, changing the world just got easier.

(Bedell 2005: 9)

Development through leverage, using lobbying and campaigning to press for global policy reforms, has assumed a growing role as an NGO strategy in pursuit of poverty reduction, fuelled in part by advocacy successes and facilitated by the revolution in communications technology. Thus, development NGOs are no longer solely deliverers of 'development' in various tangible forms, but seek to engage their constituencies in 'changing the world' through public action, emphasising, as Nelson Mandela said, that 'It is not the kings and generals who make history, but the masses of the people' (in Bedell 2005: 3). In this volume, the description of the growth in commitment to advocacy by Australian NGOs illustrated this trend, as did the increasing allocations of time and resources made to advocacy by international NGOs. Only relatively small percentages of total NGO incomes were devoted to advocacy prior to 1996. Although advocacy still accounts for a small percentage of total expenditure, more recent data suggest growing financial commitment to advocacy, with NGOs having increasingly developed and articulated advocacy strategies.

The picture of growing NGO advocacy efforts which has emerged from the pages of this volume illustrates that the ubiquity, diversity and complexity of the NGO community are mirrored in a kaleidoscope of NGO advocacy strategies and relationship styles with the many targets of their advocacy, shaped in part by the 'space' in which interactions occur, whether local, national, regional or global. The '. . . social space actors have, or lack, for enabling their ideas and projects' is of great consequence and 'How actors "expand" their room for manoeuvre is important and . . . is called effective agency' (Hilhorst 2003: 214).

Advocacy in context

The historical account of the movement of Australian NGOs into advocacy and the subsequent discussion about emerging advocacy movements at the regional level in the Mekong both demonstrate the importance of context in defining and shaping advocacy at the national and regional level. While increasing advocacy by development NGOs is a significant global trend, the content of its message is partially shaped by local and regional concerns. In Australia, and in the Mekong, advocacy about the environment brought together diverse groups and social forces in the 1990s; the issue appealed to people of all standing in society and provided, as Hirsch argued, 'an inclusive and legitimising discourse'. While the global advocacy of networks of Northern NGOs has since coalesced around global issues such as debt relief and fair trade, NGOs within South-East Asia have continued to concentrate on environmental concerns which have become part of the NGO identity, although in recent years this has been extended to include advocacy about livelihood issues and HIV/AIDS prevention. Opposition to large dams and forestry projects remains a key focus of the advocacy of local NGOs in the Mekong Region, and these issues unite in spanning social and environmental issues, but have also been the source of some friction.

The organisational form of advocacy is also shaped by the political space within which it is conducted. For Northern NGOs, often this has become global in extent, whereas the formation of NGOs is proscribed or limited in some nations in South-East Asia, as is the extent to which existing NGOs can engage in advocacy. The nature of these limitations, as Hirsch's précis of NGO advocacy in Thailand revealed, waxes and wanes with national political changes. Where little political space is afforded to advocacy, as is the case in both Vietnam and Laos, there is little room for targeted advocacy. Thus, as Hirsch and Soutar demonstrated, policy advocacy targeted directly at national governments is likely to be suppressed and NGOs have to cooperate with official bodies in order to influence policy. Even in Western democracies constraints on advocacy exist, with Ollif indicating that some Australian NGOs felt government attitudes to dissent were constraining. Recent research suggests there have been attempts by the Australian government to subdue advocacy voices using tactics which include threats of withdrawal of funding and attempts to restrict the ability of some NGOs, those deemed to engage in advocacy tangential to their main charity purpose, to offer supporters tax deductibility for donations (Maddison and Hamilton 2007: 30).

As the discussions about advocacy in the Mekong revealed, regional development plans and the major role of regional institutions in trying to realise these inevitably give development a regional focus. However, varied political environments among the nations of South-East Asia militate against strengthening the advocacy voice through regional coordination of NGO lobbying and campaigning. Moreover, with NGO activity proscribed or limited in some nations, it has been easier for Northern NGOs to target regional institutions and their development projects rather than national governments. So, apart from some generalised protests against institutions, such as those against the Asian Development Bank, advocacy has focused on megaprojects, particularly large dam developments. Even despite constraints on their activities in some nations, NGO advocacy has resulted in the implementation of safeguards for local communities and environments, and commitment of considerable resources to mitigation of, and compensation for, the effects of large-scale development projects.

Strategies for advocacy success

Successful advocacy has consisted of multiple strategies. In the Mekong, advocacy was directed both at individual projects, calling for resolution of problems arising from them, and at the institutional level, calling for reform of policies, procedures and practices. Strategies involved public events like protests, in addition to direct representation to senior staff and stakeholders. Development institutions have also instituted formal mechanisms for consultation with civil society, including hosting consultation centres.

A new form of interaction is the engagement of NGOs in development bank-funded contract work. Mirroring the complexities of NGO engagement with corporations, which Sayer detailed, some interactions result in conflict and others in cooperation. In the case of the latter, whether in receipt of bilateral funding or working cooperatively on projects with bilaterals, multilaterals or corporations, NGOs fear co-optation and that their cooperation may be seen to legitimise the broader development practices of the funding organisations, with which they may not agree. Independence therefore remains critical for some NGOs. Others, though, seek to reform 'from within', by working with institutions, accessing decision makers and building trustful relationships. These divergent approaches result in some tension between NGOs, with some believing this weakens the overall advocacy efforts of NGOs, although others believe such tension is positive and creative and results in ongoing learning relationships. Independent, 'advocacy alone' organisations are often viewed as difficult, non-constructive and unprepared to compromise, both by other NGOs and by the multilateral and bilateral agencies they hope to influence.

The importance of constructive contributions built on sound research and accurate information as a basis for advocacy became abundantly clear in the cases described by Anderson and Soutar. Emotional concentration on a single issue without offering positive options for change was seen both by some target institutions and by some NGOs as detrimental to the broader advocacy effort, although such advocacy contributes to heightened public exposure to issues and may translate into public support for change. Thus, thorough on-the-ground research was critical in the campaign against the Theun-Hinboun Dam, as was constructive criticism based on World Commission on Dams recommendations, making it difficult for the Asian Development Bank to ignore the protests about the dam.

Anderson's review of Oxfam International's efforts to influence the World Bank's HIPC policy and policy outcomes, particularly in strengthening the link between debt relief and poverty reduction programmes in beneficiary nations, is instructive. The use of mutually reinforcing strategies, involving the release of position papers and briefings, supported by lobbying, public campaigning and media coverage, was vital for success. A number of factors of particular significance emerged, including the importance of a strong underpinning of high-quality, detailed research as a basis for development of specific country-related policy documents; the coordination of the activities of OI affiliates, and collaboration with Southern organisations and other NGO coalitions; the demonstrated importance of the lobbying, widespread public campaigning and media exposure to convince the public of the need for reform and to develop Bank and shareholder interest in OI's proposals; and, crucially, the detailed specific proposals OI provided in presenting constructive propositions for achievable change. Targeting Bank shareholders was important, and understanding of the internal Bank modus operandi was essential as a basis for careful, strategic choice of particular advocacy targets. In addition, the success of OI's advocacy was based on sustained, long-term commitment to debt relief. From 1996, OI was active about reforms related to debt relief and, in 2005, despite significant gains resulting from widespread lobbying, was still pressing for an extension of debt relief following the G8's proposal to cancel the World Bank and multilateral debt of all heavily indebted poor countries.

However, taking a strong stand and working constructively with the World Bank at the policy level exposed Oxfam International to criticism by other NGOs. OI created tangible links between debt relief and social sector spending, which enabled the concept of debt relief to be sold to the sceptics, though not without accusations from other NGOs that this was establishing another form of structural adjustment. Policy proposals which seemed feasible and encouraged a gradual shift in World Bank policy were also construed by some as 'consorting with the enemy'.

As Sayer revealed, the development of policy by NGOs about their relationships with the corporate sector is nascent. However, the new, powerful mix of building relationships of trust and a sense of partnership, maintaining a degree of independence to enable a critical approach, while basing that approach on sound research and a good analytical background and providing a credible policy alternative, is also vital if NGOs are to influence corporations, as it has been for other successful NGO advocacy. This strategic combination is pragmatic and solution oriented, and recognises that some compromise may be important.

That people in their tens of thousands have stood outside G7 and G8 and many other global meetings has convinced stakeholders that there is political mileage in acting on issues, underlining the importance of public campaigning and associated strategic use of the media. The activities of other NGOs, and particularly of global networks working in alliance or coalitions as part of cohesive, coordinated networks with a strong 'brand', like Jubilee 2000, helped to create the political space within which global actors, like development banks, corporations, multilateral organisations and national governments, are receptive to suggestions for policy reform in the interests of poverty reduction.

Co-optation or cooperation?

An abiding concern in much literature and in media accounts of NGO activity is that increasing dependence on government funding may compromise NGOs' independence as critics of government policy. There is little evidence to justify concerns that receipt of government funding may inhibit the commitment of NGOs to advocacy work. Anderson could find no correlation between receipt of government funding and the advocacy expenditures of international NGOs and, among Australian NGOs, Ollif similarly could find little evidence of constraints on advocacy related to receipt of government funding. The decision about whether to advocate was clearly independent of government funding, and many NGOs refuse to accept government funding as 'a conscious decision to avoid being compromised' (Maddison and Hamilton 2007: 30). It is possible, though, that the content of advocacy and the advocacy strategies employed may be influenced not only by dependence on government funding but also by the make-up of an individual NGO's constituency; these may similarly be influenced by the relationship with, and extent of cooperation between, NGOs and their advocacy targets.

Risk compared to reward thus remains an issue for NGOs, a theme which recurs in Sayer's discussion of the risks in being seen to compromise their independence as advocates through closer engagement with those corporations whose policies and activities they seek to influence. While Sayer was painstaking in his efforts to suggest the risks are minimal and NGOs are careful to avoid compromise, the danger of co-optation – being seen to be allied with the targets of advocacy rather than maintaining distance as a basis for critical advocacy – thus remains a key issue facing NGOs in collaborating in policy development or in programmes and other forms of cooperation with corporations, multilaterals or national governments.

Networks and alliances

Throughout the book, a dominant theme has been the vital importance of strategic alliances. In facilitating alliances and networking, coordinating bodies have, without doubt, been of pivotal importance. Within Australia, the Australian Council for Overseas Aid (now the Australian Council for International Development) played a vital role in assisting Australian NGOs to strengthen and coordinate their advocacy efforts, and, in a very different context, the NGO-Forum Cambodia, as a coordinating body, assisted in the development of national and local lobbying skills, thus strengthening NGO lobbying in Cambodia, as happened in Australia.

International alliances, including with Southern organisations, have been of growing importance to NGOs' global advocacy efforts, offering shared experiences, expertise and resources, and greater impact and efficiency of campaigns. The effectiveness of building global alliances was evident in the formation of Oxfam International and its Washington Advocacy Office, the latter coordinating OI advocacy activities and being strategically located to maximise their impacts. This major strategic step demonstrated the commitment of one NGO to scaling up its advocacy, and the recognition that strength in advocacy is generated through collaboration within global networks, whether among its own members or with diverse allies. International linkages, though, can be fraught for local NGOs in nations where NGO activity is circumscribed, yet in the Mekong, where local civil society activity is limited, seemingly paradoxically international NGOs dominate regional-level advocacy. Thus, as Soutar demonstrated, the activities of Northern NGOs in challenging current development practice and articulating alternatives provide a crucial alternative to mainstream development processes.

Accountability and legitimacy

Questions about the legitimacy of NGOs to speak on behalf of Southern communities, regions and nations constitute a continuing challenge, with the majority of NGOs selecting issues for advocacy campaigns without consultation with their Southern partners. Criticisms of NGOs' failure to build effective partnerships with Southern NGOs, and to base their advocacy on in-field experiences and on those of their Southern partners, thus continue to have validity (Eade 2002; Hilhorst 2003). Advocacy-only NGOs, in particular, are often regarded as lacking the accountability and legitimacy that working relationships with Southern communities or NGOs give to organisations involved in project implementation.

Banks and borrowing governments have mixed feelings about working with NGOs and whether they should be incorporated into the development process, concerned that, when NGO activism derails a project, the NGOs are not responsible for the costs to those who lose. Thus, NGOs are often seen as 'merely shrill, less than accountable critics of development aid ready to launch campaigns against official policy' (Lewis and Opoku-Mensah 2006: 668). Despite this, recognising that dealings with them are inevitable and that often NGO action is based on sound research, development banks have been responsive to NGOs, sometimes formalising mechanisms for consultation. However, attempts by banks to be responsive, as was the Asian Development Bank in setting up a department for interaction with NGOs, have often been viewed with cynicism by NGOs, with some seeing formal consultation processes as limiting NGO access to bank staff and as an alternative to mainstreaming NGO concerns. Refusal by NGOs to negotiate through these new avenues for engagement, though, leaves them open to accusations that they are not open to compromise, particularly those campaigning NGOs which 'almost have to be radical' (Mallaby 2004: 55). Failure to accept overtures from the institutions they criticise leaves NGOs open to further accusations that they too are not accountable.

The campaign against the Theun-Hinboun Dam was initiated by a US-based advocacy organisation and supported by a network of international environmental and development NGOs, based within the Mekong and beyond. Pressure from the NGO lobby resulted in admissions about the negative impacts of the dam development. However, the slowness of the Bank to respond and the attempt to shift the burden of responsibility for mitigation and compensation to the private companies involved and to the Lao national government raise more critical questions about accountability and about who the targets of NGO advocacy should be. As Hirsch outlined, institutions such as development banks, rather than national governments or corporations, have tended to be the target of advocacy, and Sayer's work confirmed that the development of policies about engagement with the corporate sector is still in its infancy. However, in the case of the Theun-Hinboun Dam, the corporate operator of the dam assumed responsibility and did make a fruitful effort to mitigate problems for some whose livelihoods were affected and, for a short time, it seemed that the NGO heading the advocacy campaign would work with the Theun-Hinboun Hydropower Company in mitigation measures. However, that the Bank responded at all to NGO concerns indicated that it did accept a measure of responsibility for project outcomes.

NGOs targeted the ADB because their perception was that they would have greater leverage there – donor governments in Northern nations are answerable to their taxpayers and, if NGOs could influence them, they in turn could pressure the Bank to take action. Targeting the Theun-Hinboun, it was hoped, would have a wider impact than the single project and influence the Bank's consideration of future hydropower developments in the Mekong and related environmental and social issues.

For Northern NGOs, though, without links to an NGO base in Laos, legitimacy was an issue, so targeting the government was problematic. Given the polity in Laos, the campaign against the Theun-Hinboun Dam was unable to draw on legal mechanisms to assist affected citizens. Lao national, or even Laos-based international, NGOs could not speak out or easily provide information. Consequently, those on whose behalf NGOs advocated often had little knowledge of actions taken on their behalf, lending some credence to claims that NGO accountability to the poor is often 'murky' and that international NGOs' advocacy is frequently on behalf of small and non-representative local groups (Mallaby 2004). While the lack of contact with civil society groups or local people in Laos raises questions of legitimacy, advocacy NGOs see their role as essential in nations where civil society is constrained. Thus, there is potential for NGO influence in less democratic circumstances than those where coalitions and networks between international NGOs and local-level actors can be built. The case also illustrates the complexity of accountability issues. The ADB claims it is accountable to governments not NGOs, but the non-elected government in Laos is not responsive or accountable to the people of its nation. Yet the government had no responsibility for ongoing monitoring and mitigation of the negative impacts of the dam, so Bank claims that lending is legitimised by governments are curious. Expressing allied concerns, even more dramatically, as a result of NGO activism the World Bank has sometimes pulled out of projects, but national governments proceeded with them anyway, often without the environmental and social safeguards which the Bank would have imposed. Thus, it is argued, NGO activism can, in fact, harm the poor (Mallaby 2004). NGOs need to be wary that successes at one level do not exacerbate problems at another.

Evaluation

A recurrent theme of relevant literature and of this volume relates to the failure of NGOs to evaluate their advocacy. Ollif and Anderson revealed that NGOs have not consistently evaluated their advocacy but this is perhaps unsurprising, as the difficulties in doing so are enormous. Anderson's detailed research demonstrated the complexities involved, for, even with the long-term time commitment doctoral research afforded, evaluation of the effectiveness of advocacy was fraught, largely because disentangling the impacts of one NGO's work from those of other actors, or from evolutionary change within target organisations, and clearly ascribing responsibility for influencing change were nigh impossible. The example of Oxfam International is again instructive. Oxfam's position between the extremes of the HIPC debate certainly positioned it well, and its sound research enabled it to contribute to policy formation, so that even those reluctant to attribute change to it referred to it as one of the leaders in the debate. Sheer weight of evidence from exhaustive interviews and documentary analysis means that OI's success in influencing policy in some ways is incontrovertible. However, those most closely associated with OI's role in influencing policy recognised that it was one influence among many, and that others, including businesses and Northern donor governments, as well as other NGOs, civil society organisations and networks like Jubilee 2000, may have had as significant an influence or a more significant influence. It was similarly difficult to separate their influences from that of the World Bank president, who, since his appointment in 1995, had empowered progressive and reformist staff within the Bank itself, critical in creating a climate for change related to attempts to achieve poverty reduction through debt relief. The coalescence of factors which contribute to change thus makes attributing causation to any particular NGO extremely problematic.

Failure to evaluate could well be related also to lack of resources. This continues to be an issue for the NGOs, particularly for smaller organisations. That evaluating advocacy is problematic means it is difficult to attribute advocacy outcomes, which in turn makes it difficult to engender public support for, and therefore to increase public financial support for, advocacy. Many NGO supporters still prefer to donate to projects with obvious outcomes, such as supporting an individual child and his or her family, or provision of funds for schools, wells, sanitation facilities or farm animals. So advocacy work continues to be financially constrained in the face of donor preferences for a continued 'welfare approach'.

Despite the difficulties of conducting sound evaluations, NGOs have learnt from their advocacy experiences. So, for example, early enthusiasm for the development of codes of conduct for corporate behaviour waned with the realisation that they had not achieved their goals and that more binding national and international regulatory mechanisms may be more effective in controlling corporate conduct. Similarly, the growing emphasis on working in global alliances results from learning about 'what works'.

Formal advocacy evaluation has tended to focus on the impacts of campaigning and lobbying on the institutions, governments or corporations which are the target of advocacy. There is still little knowledge about how those policy gains translate into improvements in quality of life for those on whose behalf the NGOs are working. One of the challenges facing NGOs is to extend evaluation of their advocacy to attempting to remedy this knowledge deficit (Anderson 2003) – even more problematic than the comparatively easier task of assessing the extent to which their advocacy has resulted in policy change!

Are NGOs making poverty history?

There is, of course, little homogeneity in civil society. Even those NGOs united by their interest in 'development' as a basis for poverty alleviation are composed of organisations with diverse and sometimes conflicting interests. They continue to have divided views about who should be the targets of advocacy and whether reformation or abolition of development institutions should be their primary mandate. However, it is increasingly apparent that they can effect change.

Central to advocacy, whether it targets the largest provider of development finance, the World Bank, other institutions like the World Trade Organization or the European Union, the trade policies of economically powerful nations, corporations or regional institutions, is the shared belief that people can contribute to change. In the past two decades a relatively new phenomenon has been global organisation and networking around a diversity of core issues by development NGOs, united in their demands for stronger and greater participation of people in the decision-making processes which affect their lives. This book has contributed to the growing corpus of knowledge about the impacts of this global and local advocacy work. The importance of NGOs in influencing the development of policy has been acknowledged by the targets of their advocacy, other stakeholders in development processes, interested observers and the NGOs themselves. Once-fragmented grassroots movements, formerly concentrating on project work at the local level in disadvantaged communities, have become high-tech, focused and global. The shift in name of the Australian Community Aid Abroad – with the loss of the words 'Community' and 'Aid' – to Oxfam Australia, signifying membership of the global coalition Oxfam International, is representative of such change. NGOs have shown themselves to be capable of deliberate, coordinated globalisation. As with other interconnections which have facilitated globalisation, they have increasingly become a 'matter of association and connectivity, not space' (Axford 2004: 261) – and must be taken seriously.

There is, of course, much more for NGOs to do to achieve their collective goal of poverty eradication, while ever-fluctuating issues which impact on the disadvantaged in our world, notably the recent rise in concern over the impacts of global warming, mean that the focus and demands of advocacy will change. Issues of accountability and legitimacy remain a challenge for the future, and so too does work at the regional level directed at those national governments where advocacy, by various means, is constrained. Exploration of the depth and strength of alliances and how they contribute to advocacy successes is needed, and particularly the extent to which relationships with Southern NGOs support the advocacy agendas of Northern NGOs and 'provide a legitimizing platform for dissident and diverse voices from regions where economic and political power is lacking' (Taylor 2004: 273). Exploration of the influence of stakeholders on the content of advocacy messages, especially of stakeholders funding NGOs, be they governments, corporations or NGOs' own donor constituencies, would further contribute to understanding the complex dynamics of NGO–donor relations.

There is now undeniable evidence that advocacy 'works'. Significant influences on policy in the interests of poverty reduction have been made, although evaluating exactly how that translates into gains for the poor is a task yet to be undertaken (Anderson 2003). Implicitly, it is to be expected that advocacy would have some effect, and certainly it is unlikely that NGOs would commit more funds to it if they did not believe that to be the case. Commentary about Jubilee 2000 refers to its unprecedented success (Mayo 2005), and the growth in distribution of (indicating consumer demand for) fair trade products and adoption of fair trade products by travel agents, churches and a range of businesses and organisations demonstrate the public successes of NGO advocacy. Oxfam International, representative of a larger NGO which has 'scaled up' its advocacy through having 'globalised' by formalising a coalition of member organisations, has demonstrably influenced World Bank policy in strengthening the link between poverty reduction strategies and debt relief. Although a large part of this book has focused on one large globalised NGO coalition, by implication other large, international NGOs have the same capacity to effect change. Smaller NGOs, and those not part of a global network, have less chance of impacting on large organisations responsible for global policy making. However, through local coordinating bodies and global networks like the Global Call to Action against Poverty, they and their supporters are able to participate in such processes. Those NGOs whose activities are circumscribed by their national political context can also be effective agents of change, through concentration on local issues and cooperating with local institutions or through networking with international NGOs.

Whereas Northern NGOs once encouraged their supporters to exercise personal responsibility by sponsoring a child, or 'buying' a goat, a well, immunisation or educational supplies for a disadvantaged family or community in another nation, new forms of personal responsibility are being encouraged by NGOs. These include signing petitions about debt relief, lobbying politicians, buying fair trade produce, funding advocacy campaigns, joining demonstrations at major global decision-making forums, or wearing the signature white armband or similar 'brands' as a public indication of support for a global movement with the *raison d'être* to 'Make Poverty History'. Development NGOs are actively, and successfully, mobilising civil society, aiming to 'help you make us the generation that did it' (Bono, in Bedell 2005: 8). Time will tell whether, through them, for the ordinary citizen, 'changing the world just got easier'.

Notes

3 Speaking out: Australian NGOs as advocates

1 The research defined NGO size as: small = less than 10 employees, mid-size = 10–30 employees, large = more than 30 employees (see Ollif 2003).
2 A full description of the Program of Action of the ICPD can be found on the United Nations Population Fund website (www.unfpa.org) (accessed December 2006).
3 Over the last decade, Oxfam Australia has changed its name several times. In this chapter, Oxfam Australia is sometimes referred to as Community Aid Abroad to reflect its name at the time of the point being made.
4 'Super-NGOs' are also known as INGOs (international NGOs) and BINGOs (big international NGOs).
5 Native Title refers to the recognition of the rights of Aboriginal and Torres Strait Islander people within Australia to negotiate and fight to claim title to land, based on acknowledgement of their prior rights as the original occupants of the land.
6 See http://www.ausaid.gov.au/ngos/default.cfm for information about NGO accreditation with AusAID (accessed December 2006).
7 See http://www.ausaid.gov.au/ngos/display.cfm?sectionref=1776949223 for AusAID's 'Package of information: AusAID's NGO funding schemes, Module 4.2' (accessed December 2006).

4 Global action: international NGOs and advocacy

1 As all Oxfams gave consent to publication of the survey results and attribution of data to them, they are referred to by name. The other NGOs which consented to publication and attribution are similarly named. Where NGOs have restricted the attribution or publication of data, these constraints are observed by numbering, rather than naming, respondents.
2 The survey results have been published previously (Anderson 2000).
3 Intermon is the Spanish Oxfam, which in 2001 changed its name to Intermon Oxfam.
4 It is acknowledged that some part of expenditures which have been classified as 'non-advocacy' may be attributable to advocacy-related or capacity-building programmes of recipients of grants from the responding NGOs.
5 Examination of the third NGO strategic weakness identified by Edwards (1993), their failure to develop a credible alternative to neoliberal economic growth-oriented orthodoxies, was beyond the scope of the survey.
6 Governance in relation to OI is taken to mean responsibility for its vision, mission, strategy and policy. As such, governance involves focusing on the organisation's direction and longer-term strategic considerations, addressing policy in relation to

operations, defining the norms and values that are the basis of institutional functioning, and accepting responsibility for compliance with statutory requirements in the jurisdictions of registration and the external positioning of OI as the entity through which affiliates collaborate in pursuit of the OI objects. OI's governance involves the establishment and monitoring of structures and processes which enable the supervision of performance and ensure accountability to stakeholders – affiliates, donors, regulators, staff, programme beneficiaries and organisations which OI seeks to influence. Governance is distinguishable from management which is responsible for operationalising the organisation (Tandon, in Edwards and Hulme 1995: 42; Hudson 1995).

7 As at 1999, and substantially unchanged until early 2003, when the organisational structure was simplified by combining the four coordinating committees into a global coordinating team comprising skill-based representation from all affiliates (OI 2002a).

5 Oxfam, the World Bank and heavily indebted poor countries

1 From late 2005, Agir Ici of France was admitted to OI membership and is proposed to be renamed Oxfam France in 2008.

2 Oxfam's debt relief advocacy was one of three case studies of its World Bank-oriented advocacy. The others relate to Oxfam's efforts to influence the World Bank's Poverty Reduction Strategy Papers, as a poverty reduction framework to be prepared by beneficiary countries as a prerequisite for HIPC relief, and education policy, and are reported in Anderson 2003.

3 The term 'Human Development Window' was first used externally by Oxfam in its September 1998 position paper *Debt Relief and Poverty Reduction: Strengthening the Linkage* (OI 1998b).

4 The decision was made by the Interim and Development Committees of the World Bank and IMF, at their September 1996 meetings.

5 The Development Committee is a 'forum of the World Bank and the International Monetary Fund that facilitates intergovernmental consensus-building on development issues' (www.worldbank.org/devcommittee).

6 International Development Association of the World Bank.

7 Ann Pettifer was a former leader and spokesperson for Jubilee 2000.

8 Formerly Community Aid Abroad and then, to mid-2005, Oxfam Community Aid Abroad.

6 Confrontation, cooperation and co-optation: NGO advocacy and corporations

1 The international cosmetics company which is known for the high priority it gives to issues of environmental and social responsibility.

2 The quality management standards of the International Standards Organisation.

7 Risks and rewards: NGOs engaging the corporate sector

1 Approaches were made to the following agencies to seek interviews and documentation for the study: ActionAID (UK), Christian Aid (UK), Médecins sans Frontières (Belgium), Save the Children Alliance, Save the Children USA, Save the Children UK, Oxfam International, Oxfam America, Oxfam GB, Oxfam Hong Kong, Oxfam Community Aid Abroad Australia, CARE UK, CARE USA, World Development Movement (UK), World Vision UK, and World Vision USA. Comments made have not been attributed to particular individuals or organisations, and NGOs are only named where publicly available information is used. The complete research findings can be found in Sayer (2003).

8 Advocacy, civil society and the state in the Mekong Region

1 CUSO is a Canadian-based international cooperation agency that supports volunteers and a range of social justice issues. Formerly known as Canadian University Service Overseas, CUSO is now the stand-alone name for the organisation.

References

ABC (Australian Broadcasting Commission) (2006) 'Landmine campaign', ABC Radio. Available http://www.abc.net.au/rn/talks/brkfast/stories/s1629302.htm (accessed 15 December 2006).

ACC (Advocacy Coordinating Committee) (1996) 'Advocacy Workplan, 1996–1998', Unpublished, OI, Washington, DC.

ACC (1998) 'Advocacy Workplan, 1998–2000', Unpublished, OI, Washington, DC.

ACC (1999) 'Advocacy Workplan, 1999–2001', Unpublished, OI, Washington, DC.

ACFID (Australian Council for International Development) (2005) *Annual Report 2005*, Canberra: ACFID.

ACFID (2006) Australian Council for International Development online. Available http://www.acfid.asn/au (accessed 8 August 2006).

ACFOA (Australian Council for Overseas Aid) (1972) 'Big campaign launched', *Development News Digest*, 1: 3.

ACFOA (1973) 'News', *Development News Digest*, 1: 10–12.

ACFOA (1985) *Twenty Years of Service: 1985 Annual Report*, Canberra: ACFOA.

ACFOA (1988) *Directory of Member Organisations, 1987/88*, Canberra: ACFOA.

ACFOA (1989) *One World or . . . None: 1989 Annual Report*, Canberra: ACFOA.

ActionAid (2000) *Good Business: Evaluating the Impact of Community–Business Partnerships in India*, London: ActionAid.

ADAA (Australian Development Assistance Agency) (1975) *First Annual Report, 1974–75*, Canberra: Australian Government Publishing Service.

ADAB (Australian Development Assistance Bureau) (1980) *Development Co-operation, Key Statements, October 1975 – November 1980*, Canberra: Australian Government Publishing Service.

ADAB (1983) *ADAB/NGO Project Subsidy Scheme since 1980*, Canberra: ADAB/NGO Committee for Development Co-operation Publication.

ADB (Asian Development Bank) (1998a) *Cooperation between the Asian Development Bank and Nongovernmental Organisations*, Manila: ADB.

ADB (1998b) Report on site visit 6–9 May 1998, Loan N. 1329 Lao (SF): Theun-Hinboun Hydropower Project, Special Loan Review Mission, Manila.

ADB (1998c) Aide-memoire: Special Review Mission, 18–28 November. Loan N. 1329 Lao (SF): Theun-Hinboun Hydropower Project, Manila.

ADB (1998d) Letter from ADB to IRN, Manila.

ADB (2001) *Moving the Poverty Reduction Agenda Forward in Asia and the Pacific: The Long-Term Strategic Framework of the Asian Development Bank (2001–2015)*, Manila: ADB.

ADB (2002) *Landmark Joint Venture Brings More Power to the People – Connecting Nations Linking People: The Greater Mekong Subregion (GMS) Economic Cooperation Program*, Manila: ADB.

ADB (2003a) *ADB–Government–NGO Cooperation: A Framework for Action 2003–2005*, Manila: ADB.

ADB (2003b) *Theun-Hinboun Hydropower Project: Evaluation Highlights of 2002*, Manila: Operations Evaluation Department, ADB.

AIDAB (1990) *1989–90 Annual Report of the AIDAB/NGO Co-operation Program*, Canberra: AIDAB.

Aldaba, F., Antezana, P., Valderrama, M. and Fowler, A. (2000) 'NGO strategies beyond aid: perspectives from Central and South America and the Philippines', *Third World Quarterly*, 21: 669–83.

Ali, S. (2000) 'Shades of green: NGO coalitions, mining companies and the pursuit of negotiating power', in J. Bendell (ed.), *Terms for Endearment: Business, NGOs and Sustainable Development*, Sheffield: Greenleaf Publishing.

Alliband, G. (1983) 'The role of voluntary agencies in overseas aid', *World Review*, 22: 52–69.

Anderson, I. (2000) 'Northern NGO advocacy: perceptions, reality, and the challenge', *Development in Practice*, 10: 445–52.

Anderson, I. (2002) 'Northern NGO advocacy: perceptions, reality and the challenge', in D. Eade (ed.), *Development and Advocacy*, Oxford: Oxfam Great Britain.

Anderson, I. (2003) 'Towards global equity: Northern and international development organisations' advocacy examined through Oxfam International's World Bank poverty reduction policy influence', Unpublished Ph.D. thesis, University of New England, Armidale.

Anderson, N. (1964) 'Australia's voluntary foreign aid activities', *Australian Outlook*, 18: 127–42.

Anderson, S. and Cavanagh, J. (2000) *Top 200: The Rise of Corporate Global Power*, Washington, DC: Institute for Policy Studies.

*Angli*CORD (Anglicans Cooperating in Overseas Relief and Development) (2005) *Annual Report*, Melbourne: *Angli*CORD.

Annis, S. (1987) 'Can small-scale development be a large-scale policy? The case of Latin America', *World Development*, 15: 129–34.

APACE-VFEG (2006) APACE-VFEG online. Available http://www.apace.uts.edu.au (accessed 15 November 2006).

Archer, D. (1994) 'The changing roles of non-governmental organisations in the field of education (in the context of changing relationships with the state)', *International Journal of Educational Development*, 14: 223–32.

Arnold, S. (1988) 'Constrained crusaders? British charities and development education', *Development Policy Review*, 6: 183–209.

Ashman, D. (2000) *Promoting Corporate Citizenship in the Global South: Towards a Model of Empowered Civil Society Collaboration with Business*, IDR Report, 16, Boston, MA: Institute of Development Research.

Ashton, J. (1989) *Cambodia: Development Needs*, Canberra: ACFOA.

Asia Monitor Resource Centre (2001) 'Credibility gap between codes and conduct', *Asian Labour Update*, 37, Hong Kong: Asia Monitor Resource Centre, pp. 1–8.

AusAID (Australian Agency for International Development) (1995) *Review of the Effectiveness of NGO Programs*, Canberra: Australian Agency for International Development.

AusAID (1999) *Working with Australian NGOs: An Australian Aid Program Policy Paper*, Canberra: Commonwealth of Australia.

AusAID (2005a) 'Summary of official aid through Australian and non-Australian NGOs, 1999–2000 through to 2003–2004'. Online. Available http://www.ausaid.gov.au/publications/pdf/ngostatreport04/table1.pdf (accessed 11 August 2006).

AusAID (2005b) *Annual Report 2004–2005*, Canberra: Commonwealth of Australia.

Austin, J. (2000) *The Collaboration Challenge: How Nonprofits and Business Succeed through Strategic Alliances*, Hoboken, NJ: Jossey-Bass.

Axford, B. (2004) 'Global civil society or "networked globality": beyond the territorialist and societalist paradigm', *Globalizations*, 1: 249–64.

Bakker, K. (1999) 'The politics of hydropower: developing the Mekong', *Political Geography*, 18: 209–32.

Barnett, N. M. and Finnemore, M. (1999) 'The politics, power and pathologies of international organizations', *International Organization*, 53: 699–732.

Bebbington, A. (1997) 'New states, new NGOs? Crises and transitions among rural development NGOs in the Andean region', *World Development*, 25: 1755–65.

Bebbington, A. and Farrington, J. (1992) 'Non-government interaction in agricultural technology development', in M. Edwards and D. Hulme (eds), *Making a Difference: NGOs and Development in a Changing World*, London: Earthscan.

Bedell, G. (2005) *Makepovertyhistory: How You Can Help Defeat World Poverty in Seven Easy Steps*, London: Penguin Books.

Bendell, J. (ed.) (2000) *Terms for Endearment: Business, NGOs and Sustainable Development*, Sheffield: Greenleaf Publishing.

Bendell, J. (2005) 'In whose name? The accountability of corporate social responsibility', *Development in Practice*, 15: 375–88.

Bendell, J. and Lake, R. (2000) 'New frontiers: emerging NGO activities to strengthen transparency and accountability', in J. Bendell (ed.), *Terms for Endearment: Business, NGOs and Sustainable Development*, Sheffield: Greenleaf Publishing.

Blackburn, S. (1993) *Practical Visionaries: A Study of Community Aid Abroad*, Melbourne: Melbourne University Press.

Blowfield, M. (1999) *Coherence and Divergence: The Advantages and Disadvantages of Separating Social and Environmental Issues in Developing Standards and Codes of Practice for Agriculture*, Natural Resources and Ethical Trade Working Paper 5, Chatham: Natural Resources Institute.

BOND (British Overseas NGOs for Development) (1999) *The BOND Report: NGO Futures – Partnerships with the Private Sector*. Online. Available http://www.mailbase.ac.uk/links/business-ngo-relations/files/bondreport.html (accessed 17 April 2002).

Booth, D. (1985) 'Marxism and development sociology: interpreting the impasse', *World Development*, 13: 761–87.

Bray, J. (2000) 'Web wars: NGOs, companies and governments in an internet-connected world', in J. Bendell (ed.), *Terms for Endearment: Business, NGOs and Sustainable Development*, Sheffield: Greenleaf Publishing.

British Petroleum (BP) (2001) *Environmental and Social Report 2000*, London: BP Amoco.

Burbury, R. (2000) 'NGOs "slaughter" business', *Australian Financial Review*, 3 November.

Burnell, P. (1991) *Charity, Politics and the Third World*, London: Harvester Wheatsheaf.

Burns, R. (1981) 'The role of NGOs in educating Australians about North–South issues', *Development Dossier*, 6: 33–9.

Burns, R. B. (1998) *Introduction to Research Methods*, 3rd edition, Melbourne: Addison Wesley Longman.

Button, J. (2005) 'Pop politicians may not be in tune with G8', *Sydney Morning Herald*, 8 July: 11.

Bysouth, K. (1986) 'The non-governmental organisations', in P. Eldridge, D. Forbes and K. Porter (eds), *Australian Overseas Aid: Future Directions*, Sydney: Croom Helm.

CAA (Community Aid Abroad) (2000a) *Annual Report 1999–2000*, Melbourne: CAA.

CAA (2000b) *Horizons*, 9, Melbourne: CAA.

CARE (Cooperative for Assistance and Relief Everywhere) and WWF (World Wildlife Fund for Nature) (2002) *Social and Environmental Justice: Rural Poverty Eradication and Natural Resource Conservation*, Atlanta, GA: CARE and WWF.

CDCAC (Canadian Democracy and Corporate Accountability Commission) (2002) *The New Balance Sheet: Corporate Profits and Responsibility in the 21st Century*, Ottawa: CDCAC.

Chambers, R. (1997) *Whose Reality Counts? Putting the First Last*, London: IT Publications.

Chapman, J. and Fisher, T. (2000) 'The effectiveness of NGO campaigning: lessons from practice', *Development in Practice*, 10: 151–65.

Chapman, J. and Fisher, T. (2002) 'The effectiveness of NGO campaigning', in D. Eade (ed.), *Development and Advocacy*, Oxford: Oxfam Great Britain.

Chong, T., Gomez, J. and Lyons, L. (eds) (2005) 'Democracy and civil society: NGO politics in Singapore', *Sojourn Journal of Social Issues in Southeast Asia*, Special Issue, 20.

Clark, J. (1991) *Democratising Development*, London: Earthscan.

Clark, J. (1992) 'Policy influence, lobbying and advocacy', in M. Edwards and D. Hulme (eds), *Making a Difference: NGOs and Development in a Changing World*, London: Earthscan.

Clark, J. (2003) *Worlds Apart: Civil Society and the Battle for Ethical Globalization*, London: Earthscan.

Clinton, B. (2006) 'Power to the people: now everyone can make a difference', *Sydney Morning Herald*, 24 February: 13.

Coates, B. (2000) 'Getting into bed with business – some precautions', *Alliance*, 5, West Malling: Charities Aid Foundation.

Coates, B. and David, R. (2002) 'Learning for change: the art of assessing the impact of advocacy work', *Development in Practice*, 12: 530–41.

Commins, S. (1997) 'World Vision International and donors: too close for comfort?', in D. Hulme and M. Edwards (eds), *NGOs, States and Donors: Too Close for Comfort?*, New York: St Martin's Press and Save the Children Fund.

Connell, J. (1988) *Nearest Neighbour Analysis: Australian Geography and Development*, Occasional Paper 2, Sydney: University of Sydney, Research Institute of Asia and the Pacific.

Corbridge, S. (1990) 'Post-Marxism and development studies: beyond the impasse', *World Development*, 18: 623–40.

Corbridge, S. (1992) 'Third World development', *Progress in Human Geography*, 16: 584–95.

Covey, J. and Brown, D. (2001) *Critical Cooperation: An Alternative Form of Civil Society–Business Engagement*, IDR Report, 17, Boston, MA: Institute of Development Research.

Craig, D. and Porter, D. (2006) *Development beyond Neoliberalism? Governance, Poverty Reduction and Political Economy*, London and New York: Routledge.

Crane, A. (2000) 'Culture clash and mediation: exploring the cultural dynamics of business–NGO collaboration', in J. Bendell (ed.), *Terms for Endearment: Business, NGOs and Sustainable Development*, Sheffield: Greenleaf Publishing.

Currah, K. (2000) 'Putting TNC–NGO partnerships into a civil society context', in *Buy In or Sell Out? Understanding Business–NGO Partnerships*, Discussion Paper 10, Milton Keynes: World Vision UK.

Danaher, K. (ed.) (1994) *Fifty Years Is Enough*, Boston, MA: South End Press.

Dasgupta, S., Laplante, B. and Mamingi, N. (1998) *Capital Market Responses to Environmental Performance in Developing Countries*, Washington, DC: World Bank Development Research Group.

Davies, R. (2001) *Evaluating the Effectiveness of DFID's Influence with Multilaterals Part A: A Review of NGO Approaches to the Evaluation of Advocacy Work*. Online. Available http://www.mande.co.uk/docs/EEDIreport.doc (accessed 12 June 2003).

Dhanarajan, S. (2000) 'Symbiotic or parasitic: defining the relationship between TNCs and women workers', Paper presented to the ESRC Seminar Series, Harris Manchester College, Oxford, 6–7 July.

Dhanarajan, S. (2005) 'Managing ethical standards', *Development in Practice*, 15: 529–38.

Dixson, C. and Smith, D. W. (eds) (1997) *Uneven Development in South East Asia*, Aldershot: Ashgate.

Donovan, F. (1977) *Voluntary Organisations: A Case Study*, Bundoora, Victoria: Preston Institute of Technology Press.

Draper, S. (2002) 'Good work: employees as drivers and demonstrators of CSR', in R. Cowe (ed.), *No Scruples: Managing to Be Responsible in a Turbulent World*, London: Spiro Press.

Durning, A. (1989) *Action at the Grassroots: Fighting Poverty and Environmental Decline*, Worldwatch Paper 88, Washington, DC: Worldwatch Institute.

Eade, D. (2002) 'Preface', in D. Eade (ed.), *Development and Advocacy*, Oxford: Oxfam.

Eade, D. and Sayer, J. (2006) *Development and the Private Sector: Consuming Interests*, Bloomfield, CT: Kumarian Press.

Edwards, M. (1993) 'Does the doormat influence the boot? Critical thoughts on UK NGOs and international advocacy', *Development in Practice*, 3: 163–75.

Edwards, M. (1999) *Future Positive: International Cooperation in the 21st Century*, London: Earthscan.

Edwards, M. (2002) 'Organizational learning in non-governmental organizations', in M. Edwards and A. Fowler (eds), *NGO Management*, London: Earthscan.

Edwards, M. (2004) *Future Positive: International Co-operation in the 21st Century*, revised edition, London: Earthscan.

Edwards, M. and Hulme, D. (eds) (1992) *Making a Difference: NGOs and Development in a Changing World*, London: Earthscan.

Edwards, M. and Hulme, D. (1995) *Non-Governmental Organisations: Performance and Accountability – Beyond the Magic Bullet*, London: Earthscan.

Edwards, M. and Hulme, D. (1996) *Beyond the Magic Bullet: NGO Performance and Accountability in the Post-Cold War Period*, Bloomfield, CT: Kumarian Press.

Edwards, M. and Hulme, D. (2000) 'Scaling up NGO impact on development: learning from experience', in D. Eade (ed.), *Development, NGOs and Civil Society*, Oxford: Oxfam Great Britain.

Edwards, M. and Gaventa, J. (eds) (2001) *Global Citizen Action*, London: Earthscan.

Edwards, M., Hulme, D. and Wallace, T. (1999) 'NGOs in a global future: marrying local delivery to worldwide leverage', *Public Administration and Development*, 19: 117–36.

Eldridge, P. (1985) 'The Jackson Report on Australia's overseas aid program: political options and prospects', *Australian Outlook*, 39: 23–32.

Elkington, J. (1997) *Cannibals with Forks: The Triple Bottom Line of 21st Century Business*, Oxford: Capstone.

Elkington, J. and Fennell, S. (1998) 'Can business leaders satisfy the triple bottom line?', in *Visions of Ethical Business*, London: Financial Times Management.

Elkington, J. and Fennell, S. (2000) 'Partners for sustainability', in J. Bendell (ed.), *Terms for Endearment: Business, NGOs and Sustainable Development*, Sheffield: Greenleaf Publishing.

Elliott, L. (2005) 'G8 over, but it's the same old song', *Sydney Morning Herald*, 11 July: 5.

Enderle, G. and Peters, G. (1998) *A Strange Affair? The Emerging Relationship between NGOs and Transnational Companies*, London: PricewaterhouseCoopers.

Felisa, E. (2004) 'The politics of engagement: gains and challenges of the NGO coalition in Cebu City', *Environment and Urbanization*, 16: 79–93.

Ferguson, C. (1998) *A Review of UK Company Codes of Conduct*, London: Department for International Development.

FIVAS (1996) *More Water, More Fish?*, Oslo: FIVAS.

Forbes, D. (1985) 'Growth with equity equals development: a review of the Jackson Report', *Australian Geographer*, 16: 233–5.

Fortune (1999) 'Charter: responsible business in the 21st century', *Fortune*, Special advertising section, New York.

Foweraker, J. (2001) 'Grassroots movements and political activism in Latin America: a critical comparison of Chile and Brazil', *Journal of Latin American Studies*, 33: 839–65.

Fowler, A. (1991) 'The role of NGOs in changing state–society relations: perspective from Eastern and Southern Africa', *Development Policy Review*, 9: 53–84.

Fowler, A. (1995) 'Assessing NGO performance: difficulties, dilemmas and a way ahead', in M. Edwards and D. Hulme (eds), *NGOs: Performance and Accountability – Beyond the Magic Bullet*, London: Earthscan.

Fowler, A. (2000a) 'NGDOs as a moment in history: beyond aid to social entrepreneurship or civic innovation?', *Third World Quarterly*, 21: 637–54.

Fowler, A. (2000b) 'NGO futures: beyond aid – NGDO values and the fourth position', *Third World Quarterly*, 21: 589–603.

Fowler, A. and Biekart, K. (1996) 'Do private aid agencies really make a difference?' in D. Sogge (ed.), *Compassion and Calculation: The Business of Private Foreign Aid*, London: Pluto Press.

Fowler, P. and Heap, S. (2000) 'Bridging troubled waters: the Marine Stewardship Council', in J. Bendell (ed.), *Terms for Endearment: Business, NGOs and Sustainable Development*, Sheffield: Greenleaf Publishing.

Fox, J. A. and Brown, L. D. (eds) (1998) *The Struggle for Accountability: The World Bank, NGOs, and Grassroots Movements*, Cambridge, MA: MIT Press.

Frame, B. (2005) 'Corporate social responsibility: a challenge for the donor community', *Development in Practice*, 15: 422–31.

Frankfort-Nachmias, C. and Nachmias, D. (2000) *Research Methods in the Social Sciences*, 6th edition, New York: Worth Publishers.

Fukuyama, F. (1995) *Trust: The Social Virtues and the Creation of Prosperity*, London: Hamish Hamilton.

Garbutt, A. (2003) 'Civil society strengthening in Central Asia', in B. Pratt (ed.), *Changing Expectations? The Concept and Practice of Civil Society in International Development*, Oxford: INTRAC Publications.

GCAP (Global Call to Action against Poverty) (2006) Online. Available http://www.whiteband.org/ (accessed 17 January 2007).

Geldof, B. (2005) 'From Live Aid to Live 8', in M. Brown and R. Kelley (eds), *You're History: How People Make the Difference*, London and New York: Continuum Books.

George, S. and Sabelli, F. (1994) *Faith and Credit: The World Bank's Secular Empire*, Boulder, CO: Westview Press.

Giddens, A. (1982) *Sociology: A Brief, but Critical Introduction*, London: Macmillan.

Gill, I. (1997) 'Theun-Hinboun gamble pays off: hydropower project to increase Lao PDR's GDP by 7 per cent', *ADB Review*, November–December.

Glanznig, A. (1996) 'Australian aid and the ecological challenge: reflecting Rio's brave green world?', in P. Kilby (ed.), *Australia's Aid Program: Mixed Messages and Conflicting Agendas*, Melbourne: Monash Asia Institute and Community Aid Abroad.

Goldman Environmental Prize (2006) 'The power of the individual in a land of many voices'. Online. Available http://www.goldmanprize.org/node/443 (accessed 24 November 2006).

Goldsmith, E. and Hildyard, N. (1984) *The Social and Environmental Effects of Large Dams*, San Francisco, CA: Sierra Club Books.

Goldsworthy, D. (ed.) (1988) *Development Studies in Australia: Themes and Issues*, Monash Development Studies Centre Monograph 1, Melbourne: Monash University.

Gosch, E. (2005) 'Keep your cash, Geldof wants you to care', *Weekend Australian*, 2–3 July: 11.

Gray, M. (2003) 'NGOs and highland development: a case study in crafting new roles', in B. J. T. Kerkvliet, R. H. K. Heng and D. W. H. Koh (eds), *Getting Organised in Vietnam: Moving in and around the Socialist State*, Singapore: Institute of Southeast Asian Studies.

Graymore, D. and Bunn, I. (2002) *A World Summit for Business Development? The Need for Corporate Accountability in the World Summit for Sustainable Development Agenda*, London: Christian Aid. Online. Available http://www.christian-aid.org.uk/indepth/0208wssd/index.htm (accessed 23 October 2002).

Greenpeace (2002) 'Traditional adversaries call for action on climate change', Press release, 28 August. Online. Available http://www.greenpeace.org/news/details?campaign%5fid=4003&news%5fid=24688 (accessed 2 December 2002).

Greenpeace International (2006) Online. Available http://www.greenpeace.org/international/about/history (accessed 28 February 2006).

Hailey, J. (2000) 'Indicators of identity: NGOs and the strategic imperative of assessing core values', *Development in Practice*, 10: 402–7.

Hanmer, L., Healy, J. and Naschold, F. (2000) *Will Growth Halve Global Poverty by 2015?*, ODI Poverty Briefing, 8 July, London: Overseas Development Institute.

Haque, M. S. (2002) 'The changing balance of power between the government and NGOs in Bangladesh', *International Political Science Review*, 23: 411–25.

Hartnell, C. (2000) 'Will Day interview', *Alliance*, 5, West Malling: Charities Aid Foundation.

Hayes, B. and Walker, B. (2005) 'Corporate responsibility or core competence?', *Development in Practice*, 15: 405–12.

Heap, S. (1998) *NGOs and the Private Sector: Potential for Partnerships?*, Oxford: INTRAC.

Heap, S. (2000) *NGOs Engaging with Business*, Oxford: INTRAC.

Henderson, J. (2000) 'Dissonance or dialogue: changing relations with the corporate sector', *Development in Practice*, 10: 371–6.

Henry, R. (1970) 'A study of the voluntary aid movement in Australia', Unpublished MA thesis, Department of Politics, La Trobe University, Melbourne.

Hilhorst, D. (2003) *The Real World of NGOs*, London and New York: Zed Books.

Hill, H. (1980a) 'The NGOs and East Timor', *Development Dossier*, 1: 8–14.

Hill, H. (1980b) 'Australian non-governmental organizations and the Third World', *Ideas and Actions*, 137: 19–24.

Hilton, S. (2002) 'The corporatist manifesto', *Financial Times*, London, 22 April.

Hilton, S. and Gibbons, G. (2002) *Good Business: Your World Needs You*, London: Texere Publishing.

Hirsch, P. (1993) *Political Economy of the Environment in Thailand*, Manila: Journal of Contemporary Asia Publishers.

Hirsch, P. (1998) 'Community forestry revisited: messages from the periphery', in M. Victor, C. Lang and J. Bornemeier (eds), *Community Forestry at a Crossroads: Reflections and Future Directions in the Development of Community Forestry*, Bangkok: Regional Community Forestry Training Centre.

Hirsch, P. (2001) 'Globalisation, regionalisation and local voices: the Asian Development Bank and re-scaled politics of environment in the Mekong Region', *Singapore Journal of Tropical Geography*, 22: 237–51.

Hirsch, P. (2002) 'Global norms, local compliance and the human rights–environment nexus: a case study of the Nam Theun II Dam in Laos', in L. Zarsky (ed.), *Human Rights and the Environment: Conflicts and Norms in a Globalising World*, London: Earthscan.

Hirsch, P. and Wyatt, A. (2004) 'Negotiating local livelihoods: scales of conflict in the Se San River Basin', *Asia Pacific Viewpoint*, 45: 51–68.

Hobbs, J. (2001) 'From the Executive Director', *Oxfam Australia Horizons*, 1: 4, Oxfam Australia Community Aid Abroad.

Hodkinson, S. (2005) '"They've shafted us" was the cry after aid promises', *Canberra Times*, 28 October: 15.

Hodson, R. (1992) 'Small, medium or large? The rocky road to NGO growth', in M. Edwards and D. Hulme (eds), *Making a Difference: NGOs and Development in a Changing World*, London: Earthscan.

Holland, J. and Blackburn, J. (1998) 'General introduction', in J. Holland and J. Blackburn (eds), *Whose Voice? Participatory Research and Policy Change*, London: Intermediate Technology Publications.

Hudson, A. (2000a) 'Making the connection: legitimacy claims, legitimacy chains and northern NGOs' international advocacy', in D. Lewis and T. Wallace (eds), *New Roles and Relevance: Development NGOs and the Challenge of Change*, Bloomfield, CT: Kumarian Press.

Hudson, A. (2000b) 'Changing faces: the position of advocacy and NGO identity', Unpublished article.

Hudson, A. (2001a) 'From "legitimacy" to "political responsibility" in NGOs' transnational advocacy networks?', *Global Networks: A Journal of Transnational Affairs*, 1: 331–52.

Hudson, A. (2001b) 'A bigger bang for your buck? UK NGDOs and the evaluation of advocacy', Unpublished article. Online. Available http://www.alanhudson.org.uk (accessed 25 June 2003).

Hudson, M. (1995) *Managing without Profit*, London: Penguin Books.

Hulme, D. and Edwards, M. (1997) 'Conclusion: too close to the powerful, too far from the powerless', in D. Hulme and M. Edwards (eds), *NGOs, States and Donors: Too Close for Comfort?*, New York: St Martin's Press and Save the Children Fund.

Hunt, J. (1986) 'A critical assessment of Australian official development: policy and practice', Unpublished Ph.D. thesis, Department of General Studies, University of New South Wales, Sydney.

Hutchinson, M. (2000) 'NGO engagement with the private sector on a global agenda to end poverty: a review of the issues', Background paper for the Canadian Council for International Cooperation Policy Team, Ottawa.

Hutton and Cowe (2002) 'Beyond clean-up to product stewardship: the environmental agenda', in R. Cowe (ed.), *No Scruples: Managing to Be Responsible in a Turbulent World*, London: Spiro Press, pp. 81–94.

ICVA (International Council for Voluntary Action) (1990) *Relations between Southern and Northern NGOs: Effective Partnerships for Sustainable Development*, Geneva: ICVA.

IMF (International Monetary Fund) (1999) *1999 Heavily Indebted Poor Country (HIPC) Initiative: Review and Consultation*, Washington, DC: IMF.

IRN (International Rivers Network) (1999) *An Update on the Environmental and Socio-economic Impacts of the Nam Theun Hinboun Hydroelectric Dam and Water Diversion Project in Central Laos*, Berkeley, CA: IRN.

IRN (2001) IRN letter to Asian Development Bank.

IWDA (International Women's Development Agency) (2004) *Annual Report 2003–2004*, Melbourne: IWDA.

IWDA (2006) International Women's Development Agency online. Available http://www.iwda.org.au/about/htm (accessed 13 August 2006).

Jackson, R. (1985) 'Australia's foreign aid', *Australian Outlook*, 39: 13–18.

Jordan, L. and van Tuijl, P. (2000) 'Political responsibility in transnational NGO advocacy', *World Development*, 28: 2051–65.

Jubilee 2000 (2000) *Jubilee 2000 News*, 8, August, Melbourne: TEAR.

Juniper, T. (1999) 'Planet Profit', *Guardian Weekly*, 25 November.

Kalegaonkar, A. and Brown, L. (2000) *Intersectoral Cooperation: Lessons for Practice*, IDR Report, 16, Boston, MA: Institute for Development Research.

Kaosa-ard, M. and Dore, J. (eds) (2003) *Social Challenge for the Mekong Region*, Chiang Mai: Chiang Mai University Social Research Institute.

Keane, J. (2003) *Global Civil Society?*, Cambridge: Cambridge University Press.

Kearney, N. (1999) 'Corporate codes of conduct: the privatized application of labour standards', in S. Picciotto and R. Mayne (eds), *Regulating International Business: Beyond Liberalization*, London: Macmillan.

Kelly, L. (2002) 'International advocacy: measuring performance and effectiveness', Paper presented at the 2002 Australasian Evaluation Society International Conference, October/November 2002, Wollongong.

Kerkvliet, B. J. T., Heng, R. H. K. and Koh, D. W. H. (eds) (2003) *Getting Organised in Vietnam: Moving in and around the Socialist State*, Singapore: Institute of Southeast Asian Studies.

Khagram, S. (2004) *Dams and Development: Transnational Struggles for Water and Power*, Ithaca, NY: Cornell University Press.

Klein, N. (2000) *No Logo: Taking Aim at the Brand Bullies*, London: Flamingo.

Korten, D. (1987) 'Third generation NGO strategies: a key to people-centred development', *World Development*, 15, Supplement: 145–59.

Korten, D. (1990) *Getting to the 21st Century: Voluntary Action and the Global Agenda*, Bloomfield, CT: Kumarian Press.

Korten, D. (1995) *When Corporations Rule the World*, London: Earthscan.

KPMG (2002) *Measurement, Transparency, Accountability: Stepping Stones to Sustainability*, Statement by KPMG executive at Global Reporting Initiative meeting, 31 August, Johannesburg.

Leipold, G. (2002) 'Campaigning: a fashion or the best way to change the global agenda', in D. Eade (ed.), *Development and Advocacy*, Oxford: Oxfam Great Britain.

Lewis, D. and Opoku-Mensah, P. (2006) 'Moving forward research agendas on international NGOs: theory, agency and context', *Journal of International Development*, 18: 665–75.

Lindenberg, M. and Bryant, C. (2001) *Going Global: Transforming Relief and Development NGOs*, Bloomfield, CT: Kumarian Press.

Linton, A. (2005) 'Partnering for sustainability: business–NGO alliances in the coffee industry', *Development in Practice*, 15: 600–14.

Lissner, J. (1977) *The Politics of Altruism: A Study of the Political Behaviour of Voluntary Development Agencies*, Geneva: Lutheran World Federation.

Lombardo, S. (2000) 'NGOs and the multinational corporation', Paper presented at the StrategyOne conference, New York.

Long, N. (1992) 'An actor-oriented paradigm: introduction', in N. Long and A. Long (eds), *Battlefields of Knowledge: The Interlocking of Theory and Practice in Social Research and Development*, London and New York: Routledge.

McCaughey, M. and Ayers, M. (eds) (2003) *Cyberactivism: Online Activism in Theory and Practice*, New York and London: Routledge.

McCully, P. (1996) *Silenced Rivers: The Ecology and Politics of Large Dams*, London: Atlantic Highlands.

McGuinness, P. P. (2003) 'Editorial', *Quadrant*, XLVII: 2–4.

McIlwaine, C. (1998) 'Civil society and development geography', *Progress in Human Geography*, 22: 415–24.

Maddison, S. and Hamilton, C. (2007) 'The repression of the bleeding hearts', *Sydney Morning Herald*, 26–28 January: 30.

Madon, S. (2000) *International NGOs: Networking, Information Flows and Learning*, Development Informatics Working Paper Series, Working Paper 8, Manchester: Institute for Development Policy and Management.

Maina, W. (1998) 'Kenya: the state, donors and the politics of democratization', in A. Van Rooy, *Civil Society and the Aid Industry*, London: Earthscan.

makepovertyhistory (2006) Make Poverty History online. Available http://www.makepovertyhistory.org/ (accessed 15 December 2006).

Malavisi, A. (2001) 'North–South relationship: partners or pawns?', *Development Bulletin*, 55: 54–6.

Mallaby, S. (2004) 'NGOs: fighting poverty, hurting the poor', *Foreign Policy*, September/October: 50–8.

Mallaby, S. (2005) *The World's Banker*, Sydney: University of New South Wales Press.

Mares, P. (2003) *Borderline: Australia's Response to Refugee and Asylum Seekers in the Wake of the Tampa*, Sydney: University of New South Wales Press.

Martens, D. (2006) 'NGOs in the United Nations system: evaluation theoretical approaches', *Journal of International Development*, 18: 691–700.

Mayo, M. (2005) *Global Citizens: Social Movements and the Challenge of Globalization*, London: Zed Books.

Meikle, G. (2003) *Future Active: Media Activism and the Internet*, New York and London: Routledge.

Miller, K. (2001) 'The Teflon shield', *Newsweek International*, 12 March.

Minear, L. (1987) 'The other missions of NGOs: education and advocacy', *World Development*, Supplement, 15: 201–11.

Missingham, B. (2003) *The Assembly of the Poor: From Local Struggles to National Protest Movement*, Chiang Mai: Silkworm Books.

Mohan, G. (2002) 'The disappointments of civil society: the politics of NGO intervention in northern Ghana', *Political Geography*, 21: 125–54.

Molumphy, H. (1984) *For Common Decency: The History of Foster Parents Plan, 1937–1983*, Warwick, RI: Foster Parents Plan International.

Monaghan, P. (2002) 'All clear? Developing a transparent approach to company reporting', in R. Cowe (ed.), *No Scruples: Managing to Be Responsible in a Turbulent World*, London: Spiro Press.

Monbiot, G. (1997) 'Silencing of the lambs', *Guardian Weekly*, London, 31 August.

Mori, S. (1999) 'The role of NGOs in global environmental governance', Presented at the 15th United Nations University Global Seminar 'Globalization and Human Development', Tokyo.

Morrow, S. (1997) 'The changing roles of international development non-government organisations in Australia', *ACFOA Development Issues*, 2, Canberra: ACFOA.

Murphy, D. and Bendell, J. (1997) *In the Company of Partners: Business, Environmental Groups and Sustainable Development Post Rio*, Bristol: Policy Press.

Murphy, D. and Coleman, G. (2000) 'Thinking partners: business, NGOs and the partnership concept', in J. Bendell (ed.), *Terms for Endearment: Business, NGOs and Sustainable Development*, Sheffield: Greenleaf Publishing.

Murray, D. (1994) 'From battlefield to marketplace: regional economic co-operation in the Mekong zone', *Geography*, 79: 350–3.

Nason, D. and Lewis, S. (2005) 'Grand plans headed off at the impasse', *Weekend Australian*, 17–18 September: 21.

Nelson, J. (1996) *Business as Partners in Development: Creating Wealth for Countries, Companies and Communities*, London: Prince of Wales Business Leaders Forum.

Nelson, J. (1998) 'Leadership companies of the 21st century: creating shareholder value and societal value', in *Visions of Ethical Business*, London: Financial Times Management.

Nelson, J. and Zadek, S. (2000) *Partnership Alchemy: New Social Partnerships in Europe*, Copenhagen: Copenhagen Centre.

Nelson, P. (2000a) 'Heroism and ambiguity: NGO advocacy in international policy', *Development in Practice*, 10: 478–90.

Nelson, P. (2000b) 'Whose civil society? Whose governance? Decisionmaking and practice in the new agenda at the Inter-American Development Bank and the World Bank', *Global Governance*, 6: 405–31.

Nelson, P. (2002) 'Heroism and ambiguity: NGO advocacy in international policy', in D. Eade (ed.), *Development and Advocacy*, Oxford: Oxfam Great Britain.

Newbold, Y. (2002) 'Out of sight, out of mind? Scruples in the supply chain', in R. Cowe (ed.), *No Scruples: Managing to Be Responsible in a Turbulent World*, London: Spiro Press.

Newell, F. (1972) 'Action for development starts in the rich world ... Australia', *Development News Digest*, 1: 4.

Newell, P. (2000) 'From responsibility to citizenship? Corporate accountability for development', *IDS Bulletin*, 33: 91–100.

Nguyen Manh Cuong, Dao Thi Viet Nga, Le Kim Sa, Nguyen Thu Phuong, Tran Van Ha, Nguyen Hong Anh, Le Thi Thanh Huong, Nguyen Huu Thang, Nguyen Ngoc Sinh and Nguyen Danh Truong (2006) *A Work in Progress: Study on the Impacts of Vietnam's Son La Hydropower Project*, Hanoi: VUSTA.

Nichols, A. (1987) *Equal Partners:– Issues of Mission and Partnership in the Anglican World*, Report of the Mission Agencies Conference, Brisbane (1986), Sydney: Anglican Information Office.

Nichols, P. (1990) 'Project design as development process: a case study of World Vision's participation in Australia's bilateral aid program', Unpublished draft paper.

Nokia (2006) 'Corporate giving'. Online. Available http://www.nokia.com/A4275002 (accessed January 2007).

Noone, B. (1973) 'Australian economic ties with apartheid', *Development News Digest*, 2: 6–7.

Nyamagasira, W. (2002) 'NGOs and advocacy: how well are the poor represented?', in D. Eade (ed.), *Development and Advocacy*, Oxford: Oxfam Great Britain.

O'Brien, R. (1999) 'NGOs, global civil society and global economic regulation', in S. Picciotto and R. Mayne (eds), *Regulating International Business: Beyond Liberalization*, London: Macmillan.

O'Brien, R., Goetz, A. M., Scholte, J. A. and Williams, M. (2000) *Contesting Global Governance: Multilateral Economic Institutions and Global Social Movements*, Cambridge: Cambridge University Press.

OCAA (Oxfam Community Aid Abroad) (2002a) *Oxfam Australia Horizons*, 2 (2), June, Melbourne: OCAA.

OCAA (2002b) *Annual Report 2002*, Melbourne: OCAA.

OCAA (2004) *Annual Report 2004*, Melbourne: OCAA.

OECD (Organisation for Economic Co-operation and Development) (1992) *Directory of Non-Governmental Environment and Development Organisations in OECD Member Countries: Environment and Development in the Third World*, Paris: OECD.

OECD (2002) Tables in *2002 Development Cooperation Report*, Paris: OECD.

OI (Oxfam International) (1993a) 'Proposal for establishment of international Oxfam advocacy office', Unpublished, Oxford.

OI (1993b) Extract of minutes, Unpublished, Oxford.

OI (1994a) 'International advocacy office operational and management paper', Unpublished, Oxford.

OI (1994b) 'Record of agreements: conference Hong Kong 18–20 May 1994', Unpublished, Oxford.

OI (1995) *Multilateral Debt Briefing*, Washington, DC: OI.

OI (1996) *Constitution*, Oxford: OI.

OI (1997) 'Executive directors' minutes, October 1997', Oxford: OI.

OI (1997c formerly 1997), *Poor Country Debt Relief: False Dawn or a New Hope for Poverty Reduction?*, Washington, DC: OI.

OI (1998a) *Making Debt Relief Work: A Test of Political Will*, Washington, DC: OI.

OI (1998b) *Debt Relief and Poverty Reduction: Strengthening the Linkage*, Washington, DC: OI.

OI (1998c) *The HIPC Review Paper: A Wasted Opportunity*, Washington, DC: OI.

OI (1999a) Online. Available http://www.oxfam.org/aboutoi/history (accessed 20 September 2002).

OI (1999b) Online. Available http://www.oxfam.org/aboutoi/organise (accessed 21 September 2002).

OI (1999c) Online. Available http://www.oxfam.org/aboutoi/glossary (accessed 21 September 2002).

OI (1999d) '1999–2000 budget', Unpublished, Oxford.

OI (1999e formerly 1999) *Outcome of the IMF/World Bank September 1999 Annual Meetings: Implications for Poverty Reduction and Debt Relief*, Washington, DC: OI.

OI (1999f) *Education Now! Break the Cycle of Poverty: Executive Summary*, Washington, DC: OI.

OI (2000) *HIPC Leaves Poor Countries Heavily in Debt: New Analysis*, Washington, DC: OI.

OI (2001a) 'Strategic Plan 2001–2004'. Online. Available http://www.oxfam.org/eng/about_strat (accessed 28 September 2002).

OI (2001b) *Debt Relief: Still Failing the Poor*, Washington, DC: OI.

OI (2001c) Online. Available http://www.oxfam.org/about/strategic_plan/citizens4 (accessed 28 September 2002).

OI (2002a) 'Board and executive directors brief – November 2002', Unpublished, Oxford.

OI (2002b) '2002 budget', Unpublished, Oxford.

OI (2003) '2003 budget', Unpublished, Oxford.

OI (2005) *Beyond HIPC: Debt Cancellation and the Millennium Development Goals*, Washington, DC: OI.

Ollif, C. (2003) 'Speaking up . . . the advocacy work of Australian aid and development non-governmental organisations', Unpublished Ph.D. thesis, University of New England, Armidale.

Opoku-Mensah, P. (2001) 'The rise and rise of NGOs: implications for research', *Tidsskrift ved Institutt for sosiologi og statsvitenskap*, 1: 1–3.

O'Rourke, D. (2001) 'Monitoring the monitors: a critique of PricewaterhouseCoopers' labour monitoring', *Asian Labour Update*, 37, Hong Kong: Asia Monitor Resource Centre.

Osborne, M. (2004) *River at Risk: The Mekong and the Water Politics of China and Southeast Asia*, Sydney: Lowy Institute.

Oxfam Australia (2005) 'Make Poverty History: sign the petition'. Online. Available http://www.oxfam.org.au/campaiigns/mtf/povertyhistory/petition.php (accessed August 2005).

Oxfam Great Britain (1998) 'Strategy for Oxfam GB's engagement with the private sector', Unpublished policy paper, Oxford.

Oxfam Hong Kong (2001) 'Oxfam Hong Kong and the corporate sector: engagement and change', Unpublished policy paper, Hong Kong.

Oxfam, Save the Children and VSO (2002) *Beyond Philanthropy: The Pharmaceutical Industry, Corporate Social Responsibility and the Developing World*, Oxford: Oxfam.

Oxfam United Kingdom and Ireland (1993) 'From village council to United Nations: Oxfam's strategic intent 1994–99', Unpublished, Oxford.

Pante, F. (1997) 'Investing in regional development: Asian Development Bank', in B. Stensholt (ed.), *Developing the Mekong Region*, Clayton: Monash Asia Institute.

Parnwell, M. and Bryant, R. (1996) *Environmental Change in South-East Asia: People, Politics and Sustainable Development*, London: Routledge.

Partners in Change (2002) *Strategy 2001–2005*, New Delhi. Online. Available http://www.picindia.org/strategy.htm (accessed 17 October 2002).

Paul, S. and Israel, A. (eds) (1991) *Non-Governmental Organisations and the World Bank*, Washington, DC: World Bank.

Peacock, A. (1980) *Australia's Overseas Development Assistance Program, 1980–81*, Budget Paper 8, Canberra: Australian Government Publishing Service.

Pearce, J. (2000) 'Development, NGOs and civil society: the debate and its future', in D. Eade (ed.), *Development, NGOs and Civil Society*, Oxford: Oxfam Great Britain.

Perreault, T. (2003) 'Changing places: transnational networks, ethnic politics, and community development in the Ecuadorian Amazon', *Political Geography*, 22: 61–88.

Peters, G. (1998) 'Reputation: the search engine of the future', in *Visions of Ethical Business*, London: Financial Times Management.

Phillips, M. (2006) 'Fair traders debate change of retail tactics', *Australian Financial Review*, 28 June: 76–7.

Phongpaichit, P. and Baker, C. (2004) *The Business of Politics in Thailand*, Chiang Mai: Silkworm Books.

Phongphit, S. (1986) *Back to the Roots: Village and Self-Reliance in a Thai Context*, Bangkok: Rural Development Documentation Centre.

Picciotto, S. and Mayne, R. (eds) (1999) *Regulating International Business: Beyond Liberalization*, London: Macmillan.

Plante, C. and Bendell, J. (2000) 'The art of collaboration: lessons from emerging environmental business–NGO partnerships in Asia', in J. Bendell (ed.), *Terms for Endearment: Business, NGOs and Sustainable Development*, Sheffield: Greenleaf Publishing.

Porter, D. (1990) 'Cutting stones for development: issues for Australian agencies', Unpublished paper.

Porter, D. and Clark, K. (1985) *Questioning Practice: Non-Government Aid Agencies and Project Evaluation*, Development Dossier 16, Canberra: ACFOA.

Porter, D. and Kilby, P. (1996) 'Strengthening the role of civil society? A precariously balanced answer', in P. Kilby, *Australia's Aid Program: Mixed Messages and Conflicting Agendas*, Melbourne: Monash Asia Institute and Community Aid Abroad.

Potter, R., Binns, T., Elliott, J. and Smith, D. (2004) 'Civil society, NGOs and development', in *Geographies of Development*, 2nd edition, Harlow: Pearson Education.

Prahalad, C. and Hart, S. (2002) 'The fortune at the bottom of the pyramid', *Strategy + Business*, Maclean, VA: Booz Allen Hamilton.

Proctor, D. (2000a) 'The HIV/AIDS Development Network Australia (HIDNA) as an advocacy tool within the Australian NGO community', Presented to the Public Health Association of Australia's Conference, Canberra.

Proctor, D. (2000b) 'Advocacy for reproductive health', Presented to the World Health Organization Symposium: Reproductive Health Research in Asia and the Pacific, Australian National University, Canberra.

Publish What You Pay (2002) '"Publish What You Pay" appeal signatories'. Online. Available http://www.publishwhatyoupay.org/ngos.html (accessed 19 December 2002).

Raffer, K. and Singer, H. (1996) *The Foreign Aid Business: Economic Assistance and Development Co-operation*, Cheltenham and Brookfield, MA: Edward Elgar.

Redden, J. (1999) 'Birmingham conference report', Internal ACFOA document, ACFOA, Canberra.

Reid, A. (1986) 'Red herrings and yellow brick roads: Australia's private foreign aid community, 1970–1982', Unpublished B.Litt. thesis, Department of Political Science, Australian National University, Canberra.

Reimann, K. (2003) 'Building networks from the outside in: Japanese NGOs and the Kyoto climate change conference', in J. Smith and H. Johnston (eds), *Globalization and Resistance: Transnational Dimensions of Social Movements*, Oxford: Rowman & Littlefield.

Rich, B. (1994) *Mortgaging the Earth: The World Bank, Environmental Impoverishment, and the Crisis of Development*, London: Earthscan.

Richards, R. (1981) *The Zamboanga del Sur Project:– The Impact on the Poor*, Evaluation report prepared on behalf of Community Aid Abroad, Melbourne: CAA.

Riddell, R. (1987) *Foreign Aid Reconsidered*, London: James Currey.

Rigg, J. (1997) *Southeast Asia: The Human Landscape of Modernization and Development*, London: Routledge.

Robbins, R. (2006) *Global Problems and the Culture of Capitalism*, 3rd edition, Boston, MA: Pearson Allyn & Bacon.

Roberts, S., Jones III, J. and Fröhling, O. (2005) 'NGOs and the globalization of managerialism: a research framework', *World Development*, 33: 1845–64.

Roche, C. (1994) 'It's not size that matters: ACCORD's experience in Africa', in M. Edwards and D. Hulme (eds), *Making a Difference*, London: Earthscan.

Roche, C. (1999) *Impact Assessment for Development Agencies: Learning to Value Change*, Oxford: Oxfam.

Roche, C. and Bush, A. (1997) 'Assessing the impact of advocacy work', *Appropriate Technology*, 24: 9–13.

Roddick, A. (2000) 'Foreword', in J. Bendell (ed.), *Terms for Endearment: Business, NGOs and Sustainable Development*, Sheffield: Greenleaf Publishing.

Rodgers, C. (2000) 'Making it legit: new ways of generating corporate legitimacy in a globalising world', in J. Bendell (ed.), *Terms for Endearment: Business, NGOs and Sustainable Development*, Sheffield: Greenleaf Publishing.

Rolfe, B. (2005) 'Building an electronic repertoire of contention', *Social Movement Studies*, 4: 65–74.

Roper, M. (2000) 'On the way to a better state? The role of NGOs in the planning and implementation of protected areas in Brazil', *GeoJournal*, 52: 61–9.

Ross, N. (1988) 'From the Chairman . . .', *ACFOA Annual Report*, Canberra: ACFOA.

Rotham, F. and Oliver, P. (2002) 'From local to global: the anti-dam movement in Southern Brazil, 1979–1992', in J. Smith and H. Johnston (eds), *Globalization and Resistance: Transnational Dimensions of Social Movements*, Oxford: Rowman & Littlefield.

Royo, A. G. (1998) 'Planafloro in Rondonia: the limits of leverage', in J. B. Fox and L. D. Brown (eds), *The Struggle for Accountability: The World Bank, NGOs, and Grassroots Movements*, Cambridge, MA: MIT Press.

Rugendyke, B. (1991) 'Unity in diversity: the changing face of the Australian NGO community', in L. Zivetz *et al.* (ed.), *Doing Good: The Australian NGO Community*, Sydney: Allen & Unwin.

Rugendyke, B. (1994) 'Compassion and compromise: the policy and practice of Australian non-government development aid agencies', Unpublished Ph.D. thesis, Department of Geography, University of New England, Armidale.

Rumansara, A. (1998) 'Indonesia: the struggle of the people of Kedung Ombo', in J. B. Fox and L. D. Brown (eds), *The Struggle for Accountability: The World Bank, NGOs, and Grassroots Movements*, Cambridge, MA: MIT Press.

Rundell, P. (2000) 'Partnerships with TNCs: an attempt to compartmentalize ethics?', in *Buy In or Sell Out? Understanding Business–NGO Partnerships*, Discussion Paper 10, Milton Keynes: World Vision UK.

Salmen, L. and Eaves, A. (1991) 'Interactions between non-governmental organisations, governments, and the World Bank: evidence from bank projects', in S. Paul and A. Israel (eds), *Non-Governmental Organisations and the World Bank*, Washington, DC: World Bank.

Saravanamuttu, P. (1998) 'Sri Lanka: civil society, the nation and the state-building challenge', in A. Van Rooy (ed.), *Civil Society and the Aid Industry*, London: Earthscan.

Saxby, J. (1996) 'Who owns the private aid agencies?', in D. Sogge (ed.), *Compassion and Calculation: The Business of Private Foreign Aid*, London: Pluto Press.

Sayer, J. (2003) 'Development NGOs policies on engagement with the corporate sector: an assessment of coherence', Unpublished M.Litt. in Development Studies dissertation, University of New England, Armidale.

Sayer, J. (2005) 'Do more good, do less harm: development and the private sector', *Development in Practice*, 15: 251–68.

Schwartz, P. and Gibb, B. (1999) *When Good Companies Do Bad Things: Responsibility and Risk in an Age of Globalization*, New York: John Wiley & Sons.

Sen, B. (1987) 'NGO self-evaluation', *World Development*, Supplement, 15: 161–7.

Sharp, N. (1978) 'The aid debate:– the bonds of charity', *Arena*, 50: 47–53.

Shell International (2001) *People, Planet and Profits: The Shell Report 2001*, London: Shell International.

Shigetomi, S. (ed.) (2002) *The State and NGOs: Perspective from Asia*, Singapore: Institute of Southeast Asian Studies.

Shoemaker, B. (1998) *Trouble on the Theun-Hinboun: A Field Report on the Socio-economic and Environmental Effects of Nam-Theun-Hinboun Hydropower Project in Laos*, Berkeley, CA: International Rivers Network.

Shoesmith, D. (1982) 'The politics of aid: the Northern Samar integrated rural development scheme', *Development Dossier*, 2: 30–2.

Simon, D. (2003) 'Dilemmas of development and the environment in a globalizing world: theory, policy and praxis', *Progress in Development Studies*, 3: 5–41.

Simpson, A. (2002) 'Money talks: the rise of socially responsible investors', in R. Cowe (ed.), *No Scruples? Managing to Be Responsible in a Turbulent World*, London: Spiro Press.

Sklair, L. (1988) 'Transcending the impasse: metatheory, theory and empirical research in the sociology of development and underdevelopment', *World Development*, 16: 697–709.

Slater, D. (1992) 'Theories of development and politics of the post-modern:– exploring a border zone', *Development and Change*, 23: 283–319.

Smillie, I. (1995) *The Alms Bazaar: Altruism under Fire – Non-Profit Organisations and International Development*, Ottawa: International Development Research Centre.

Smillie, I. (1999a) 'Australia', in I. Smillie and H. Helmich (eds), *Stakeholders: Government– NGO Partnerships for International Development*, London: Earthscan.

Smillie, I. (1999b) 'At sea in a sieve? Trends and issues in the relationship between Northern NGOs and Northern governments', in I. Smillie and H. Helmich (eds), *Stakeholders: Government–NGO Partnerships for International Development*, London: Earthscan.

Smillie, I. and Helmich, H. (1993) *Non-Governmental Organisations and Governments: Stakeholders for Development*, Paris: OECD.

Sogge, D. (ed.) (1996a) *Compassion and Calculation: The Business of Private Foreign Aid*, London: Pluto Press.

Sogge, D. (1996b) 'Settings and choices', in D. Sogge (ed.), *Compassion and Calculation: The Business of Private Foreign Aid*, London: Pluto Press.

Sogge, D. (1996c) 'Northern Lights', in D. Sogge (ed.), *Compassion and Calculation: The Business of Private Foreign Aid*, London: Pluto Press.

Sogge, D. and Zadek, S. (1996) '"Laws" of the market?', in D. Sogge (ed.), *Compassion and Calculation: The Business of Private Foreign Aid*, London: Pluto Press.

Sprechmann, S. and Pelton, E. (2001) *Advocacy Tools and Guidelines: Promoting Policy Change*, Atlanta, GA: CARE. Online. Available http://www.careusa.org/getinvolved/advocacy/tools.asp#english (accessed 24 October 2002).

Stent, W. (1985) 'The Jackson Report: a critical review', *Australian Outlook*, 39: 33–8.

Stephenson, C. (2000) 'NGOs and the principal organs of the United Nations', in P. Taylor and A. Groom (eds), *The United Nations at the Millennium: The Principal Organs*, London and New York: Continuum.

Stott, J. (1975) *Christian Mission in the Modern World*, London: Falcon.

Sutcliffe, H. (2001) 'Risky businesses need help', *Observer*, 8 July: 14.

Tadem, T. S. E. (2003) 'Thai social movements and the anti-ADB campaign', *Journal of Contemporary Asia*, 33: 377–98.

Taylor, P. (2004) 'NGOs in the world city network', *Globalizations*, 1: 265–77.

Tendler, J. (1982) *Turning Private Voluntary Organisations into Development Agencies: Questions for Evaluation*, Program Evaluation Discussion Paper 12, Washington, DC: US AID.

Tennyson, R. (2000) 'Business–NGO partnerships: significant paradigm shift or passing fad?', in *Buy In or Sell Out? Understanding Business–NGO Partnerships*, Discussion Paper 10, Milton Keynes: World Vision UK.

Tepper Marlin, A. (1998) 'Visions of social accountability: SA8000', in *Visions of Ethical Business*, London: Financial Times Management.

Thomas, T. and Eyres, B. (1998) 'Why an ethical business is not an altruistic business', in *Visions of Ethical Business*, London: Financial Times Management.

Tiffen, R., Busby, S., Ross, J. and Storz, M. (1977) *ACFOA Report, Stage 1, Research Report*, Melbourne: Applied Sociology Department, Caulfield Institute of Technology.

Tussie, D. and Casaburi, G. (2000) 'From global to local governance: civil society and the multilateral development banks', *Global Governance*, 6: 399–403.

Tvedt, T. (2006) 'The international aid system and the non-governmental organisations: a new research agenda', *Journal of International Development*, 18: 677–90.

UNDP (United Nations Development Programme) (1993) *Human Development Report 1993*, New York and Oxford: Oxford University Press.

UNDP (1996) *Human Development Report 1996: Economic Growth and Human Development*, Oxford: Oxford University Press.

UNDP (2001) *Human Development Report 2001: Making New Technologies Work for Human Development*, Oxford: Oxford University Press.

UNDP (2002) *Human Development Report 2002: Deepening Democracy in a Fragmented World*, Oxford and New York: Oxford University Press.

UNDP (2005) *Human Development Report 2005: International Cooperation at a Crossroads: Aid, Trade and Security in an Unequal World*, Oxford: Oxford University Press.

UNECA (United Nations Economic Commission for Africa) (1999) 'Strengthening the links between debt relief and poverty reduction: summary report of the HIPC review seminar', Addis Ababa. Online. Available http://www.worldbank.org/hipc/related-papers/addis-related_papers/ (accessed July 2003).

UNEP (United Nations Environment Programme) (1994) *Partnerships for Sustainable Development*, Paris: UNEP.

United Nations (2002) *Implementation Plan of the World Summit on Sustainable Development*, New York: United Nations. Online. Available http://www.iisd.ca/wssd/download%20files/impplan_26june.pdf (accessed 2 December 2002).

Utting, P. (2005) 'Corporate responsibility and the movement of business', *Development in Practice*, 15: 375–88.

Uvin, P. J. and Brown, D. (2000) 'Think large and act small: toward a new paradigm for NGO scaling up', *World Development*, 28: 1409–19.

Vale, A. (1985) 'Non-government organisations in the post-Jackson era', Unpublished paper prepared for the National Executive of Community Aid Abroad (CAA), Melbourne.

Vandergeest, P. (1996) 'Property rights in protected areas: obstacles to community involvement as a solution in Thailand', *Environmental Conservation*, 23: 259–68.

Van Rooy, A. (2000) 'Good news! You may be out of a job: reflections on the past and future 50 years for Northern NGOs', *Development in Practice*, 10: 300–18.

Vaux, T. (2001) *The Selfish Altruist: Relief Work in Famine and War*, London: Earthscan.

Verhelst, T. (1990) *No Life without Roots: Culture and Development*, London: Zed Books.

Vidal, J. (1999) 'Warfare across the Web', *Guardian Weekly*, 7 February: 20.

Waddell, S. (1997) *Market–Civil Society Partnership Formation: A Status Report on Activity Strategies and Tools*, IDR Report, 15, Boston, MA: Institute for Development Research.

Waddell, S. (2000) 'Complementary resources: the win–win rationale for partnerships with NGOs', in J. Bendell (ed.), *Terms for Endearment: Business, NGOs and Sustainable Development*, Sheffield: Greenleaf Publishing.

Waddell, S. and Brown, L. (1997) *Fostering Intersectoral Partnering: A Guide to Promoting Cooperation among Government, Business, and Civil Society Actors*, IDR Report, 13, Boston, MA: Institute for Development Research.

Walker, B. (2000) 'Loosing the chains', *World Vision News*, 18.

Walsh, P. (1980) 'The politics of aid to East Timor', *Development Dossier*, 1: 16–21.

Warhurst, A. (2001) 'Corporate citizenship and corporate social investment: drivers of tri-sector partnerships', *Journal of Corporate Citizenship*, 1: 57–73.

War on Want (2002a) *Declaration on Harnessing Currency Transactions to Tackle Global Poverty*, London: War on Want. Online. Available http://www.waronwant.org/?lid=1452 (accessed 12 November 2002).

War on Want (2002b) *Tobin Tax Network: Signatory Organizations*, London: War on Want. Online. Available http://www.waronwant.org/?lid=3159 (accessed 12 November 2002).

Warren, T. (1999) *A Monitoring Study to Assess the Localised Impacts Created by the Nam Theun-Hinboun Hydro-scheme on Fisheries and Fish Populations*, Vientiane: Theun-Hinboun Power Company.

Watkins, K. (1995) *The Oxfam Poverty Report*, Oxford: Oxfam United Kingdom and Ireland.

Watts, M. (1988) 'Deconstructing determinism: Marxism, development theory and a comradely critique of "capitalist world development": a critique of radical development geography', *Antipode*, 20: 142–68.

Webb, J. (1977a) 'Australian overseas aid – governmental and private', *Dyason House Papers*, January: 6–8.

Webb, J. (1977b) *Beyond Aid*, Research and Information Service 17, Canberra: ACFOA.

Weir, A. (2000) 'Meeting social and environmental objectives through partnership: the experience of Unilever', in J. Bendell (ed.), *Terms for Endearment: Business, NGOs and Sustainable Development*, Sheffield: Greenleaf Publishing.

Weiss, T. and Collins, C. (1996) *Humanitarian Challenges and Intervention: World Politics and the Dilemmas of Help*, Boulder, CO: Westview Press.

Welch, Jr, C. E. (2001) 'Conclusion', in C. E. Welch, Jr (ed.), *NGOs and Human Rights: Promise and Performance*, Philadelphia, PA: University of Pennsylvania Press.

Whelan, T. (1982) '"Tea" as a means of social change?', *Development Dossier*, 8: 28–9.

White, J. (1971) *The Poor Country's View of a Wicked World*, Institute of Development Studies Seminar Paper 18, Sussex: IDS.

Wikipedia (2006a) 'Non-governmental organization', Wikipedia Encyclopedia online. Available http://www.en.wikipedia.org/wiki/Non-governmental_organization (accessed 29 December 2006).

Wikipedia (2006b) 'Tobin tax', Wikipedia Encyclopedia online. Available http://en.wikipedia.org/wiki/Tobin_tax (accessed 29 December 2006).

Williams, A. (1990) 'A growing role for NGOs in development', *Finance and Development*, 27: 31–3.

Williams, P. (2005) 'Leveraging change in the working conditions of UK homeworkers', *Development in Practice*, 15: 546–58.

Wiseberg. L. (2001) 'The internet: one more tool in the struggle for human rights', in C. Welch, Jr (ed.), *NGOs and Human Rights: Promise and Performance*, Philadelphia, PA: University of Pennsylvania Press.

World Bank Group (1998a) *The Initiative for Heavily Indebted Poor Countries: Review and Outlook*, Washington, DC: World Bank Development Committee.

World Bank Group (1998b) *HIPC Initiative: A Progress Report*, Washington, DC: World Bank.

World Bank Group (1999a) *Heavily Indebted Poor Countries (HIPC) Initiative:– Strengthening the Link between Debt Relief and Poverty Reduction*, Washington, DC: World Bank.

World Bank Group (1999b) *The HIPC Initiative: Delivering Debt Relief to Poor Countries*, Washington, DC: World Bank.

World Bank Group (1999c) 'Canada's Chrétien calls for more debt relief under HIPC', *Development News*, 29 March – 2 April, Washington, DC: World Bank.

World Bank Group (1999d) *HIPC Review: Perspectives on the Current Framework and Options for Change*, 2 April, Washington, DC: World Bank.

World Bank Group (1999e) 'Relieving the debt of the poorest: Bank consults partners on HIPC', *Development News*, 5–9 April, Washington, DC: World Bank.

World Bank Group (1999f) *HIPC Initiative – Progress to Date: Progress Report by the Managing Director of the IMF and the President of the World Bank*, 21 April, Washington, DC: World Bank.

World Bank Group (1999g) *Modifications to the Heavily Indebted Poor Countries Initiative*, Washington, DC: World Bank.

World Bank Group (1999h) *Heavily Indebted Poor Countries (HIPC) Initiative:– Strengthening the Link between Debt Relief and Poverty Reduction*, Washington, DC: World Bank.

World Bank Group (1999i) *Building Poverty Reduction Strategies in Developing Countries*, Washington, DC: World Bank Development Committee.

World Bank Group (1999j) *Communiqué, September 27 1999*, Washington, DC: World Bank Development Committee.

World Bank Group (2000) *Human Development Report 2000/2001: Attacking Poverty*, Oxford: Oxford University Press.

World Development Movement (2002) *People before Profits*, London: World Development Movement. Online. Available http://www.wdm.org.uk/campaign/history.htm (accessed 20 December 2002).

World Vision Australia (2002) *Action News*, Autumn, Melbourne: WVA.

World Vision Australia (2004) *Annual Review*, Melbourne: WVA.

World Vision UK (2000) *Buy In or Sell Out? Understanding Business–NGO Partnerships*, Discussion Paper 10, Milton Keynes: World Vision UK.

Yanacopulos, H. (2002) 'Think local, act global: transnational networks and development', in J. Robinson (ed.), *Development and Displacement*, Oxford: Oxford University Press.

Yang, G. (2005) 'Environmental NGOs and institutional dynamics in China', *China Quarterly*, 181: 46–66.

Index